BERKELEY'S REVOLUTION IN VISION

BERKELEY'S REVOLUTION IN VISION

Margaret Atherton

Cornell University Press

ITHACA AND LONDON

First published 1990 by Cornell University Press.

International Standard Book Number 0-8014-2358-9
Library of Congress Catalog Card Number 89-49632
Printed in the United States of America.
Librarians: Library of Congress cataloging information appears on the last page of the book.

♾ The paper used in this publication meets the minimum requirements of the American National Standard for Permanence of Paper for Printed Library Materials Z39.48–1984.

To R.A.S.

Contents

Part Three: Some Implications

Acknowledgments

I have learned a great deal from previous writers on George Berkeley's *New Theory of Vision*, especially from the authors of the most recent extensive treatments, D. M. Armstrong and George Pitcher. I have in addition some particular debts: to Nelson Goodman for presenting the *New Theory* to me as an important document in the theory of vision; to C. M. Turbayne for his overall approach to Berkeley, with which I find myself in general sympathy; and especially to Robert Schwartz for his range of knowledge of Berkeley's *New Theory* and of the theory of vision and for his infinite patience in sharing this knowledge with me. I owe thanks to Margaret Wilson, Kenneth Winkler, George Pappas, Robert Schwartz, and several anonymous reviewers for their criticisms of earlier drafts. A version of Chapter 12 was read at the Central Division American Philosophical Association meetings in Chicago in 1989. I am grateful to David Kline and to the audience for their comments.

My work on this book was supported by an NSF grant # 8606587 and by a sabbatical from the University of Wisconsin–Milwaukee.

M. A.

Milwaukee, Wisconsin

Abbreviations

Works	*The Works of George Berkeley, Bishop of Cloyne*, 9 vols., ed. A. A. Luce and T. E. Jessop (Edinburgh: Thomas Nelson, 1948–57). References, except to the works listed below, are by volume and page number.
A	*Alciphron or the Minute Philosopher* (1731) (*Works*, vol. 3, 1950). References are by part and section number and by page number.
NTV	*An Essay Towards a New Theory of Vision* (1709) (*Works*, vol. 1, 1948). References are by section numbers.
PC	*Philosophical Commentaries* (1707–1708) (*Works*, vol. 1, 1948). References are by entry numbers (Notebook B = entries 1–399; Notebook A = entries from 400 onward).
PHK	*A Treatise concerning the Principles of Human Knowledge*, Part I (1710) (*Works*, vol. 2, 1949). References are by section numbers.
PHK, Intro.	Introduction to *The Principles of Human Knowledge* (1708) (*Works*, vol. 2, 1949). References are by section numbers.
3D	*Three Dialogues between Hylas and Philonous* (1713) (*Works*, vol. 2, 1949). References are by dialogue number and page number.
ST	Nicolas Malebranche, *The Search after Truth* (1674–1675), translated by Thomas M. Lennon and Paul J. Olscamp (Columbus: Ohio State University Press, 1980). References are by book, chapter and section numbers.
TVV	*Theory of Vision Vindicated and Explained* (1733) (*Works*, vol 1, 1948). References are by section numbers.

PART ONE

The Historical Background to Berkeley's Project

1

Introduction

In 1709 George Berkeley published his first substantial work, *An Essay towards a New Theory of Vision*. As a contribution to the theory of vision, this work was undeniably successful. Berkeley proposed a particular approach to the question how perceivers see. According to him, perception requires learning; the way we perceive the visual world is not a matter of direct sensory stimulation but reflects what, in our experience, has been associated with visual cues or stimuli. In the *New Theory*, Berkeley contrasted his approach to vision with another, which he called the geometric.[1] According to the geometric theory, the visual system operates by means of principles that allowed it to calculate a visual percept from the immediate sensory data. On this account, we don't learn to see; rather, what we see is the result of and depends upon necessary connections holding between the way in which we are visually stimulated and the visual world we end up seeing. After Berkeley wrote the *New Theory*, use of a geometric paradigm to solve problems in perception became far less prevalent, and research conducted according to the learning paradigm Berkeley had introduced became increasingly common. It is fair to say that Berkeley's *New Theory of Vision* constituted a genuinely revolutionary turn in the history of the theory of vision.

Berkeley's revolution, moreover, had remarkably long-lasting

1. It is somewhat complicated to determine the exact nature of the geometric theory as a historical entity. I discuss this matter in some detail below; see Chapters 2 and 3.

consequences. Writing in 1842, John Stuart Mill claimed that Berkeley's theory of vision "had remained, almost from its first promulgation, one of the least disputed doctrines in the most disputed and most disputable of all sciences, the Science of Man."[2] More than one hundred years later, in 1950, when J. J. Gibson characterized the received view against which he intended to argue, the theory he described continued to fall within the Berkeleian paradigm:

> The accepted view of perception is still that the percept is never completely determined by the physical stimulus. Instead, the percept is something essentially subjective in that it depends on some contribution made by the observer himself. Perception goes beyond the stimuli and is superposed on sensations. The sensations are basic, and, being parts of our organic equipment, tend to be the same for all. Perceptions, however, are secondary and, depending on the peculiarities and past experience of the individual, may vary from one observer to another.[3]

Nor has Berkeley's basic approach been entirely abandoned, even though the different theories put forward by Gibson and others make Mill's claim no longer true. It continues to shape at least some work in perception.[4]

Since the time of Mill, however, philosophers have not always treated Berkeley's *New Theory* with the respect such a revolutionary document deserves. There has not been much attempt to explore what it is about Berkeley's theory that made it so successful. Instead, it is far more common to find philosophers treating the *New Theory* as essentially misguided, on the wrong track. The reason is that many philosphers take Berkeley's wider projects, as expressed in his two best-known works, *A Treatise concerning the Principles of Human Knowledge* (1710) and *Three Dialogues between Hylas and Philonous*

2. John Stuart Mill, "Bailey on Berkeley's Theory of Vision," in *Dissertations and Discussions*, vol. 2 (New York, 1973), p. 84.

3. J. J. Gibson, *The Perception of the Visual World* (Boston, 1950). I am grateful to Robert Schwartz for calling my attention to this passage.

4. Irvin Rock, in a standard textbook on perception, says of Berkeley's theory, "This way of looking at perception had a profound effect on all subsequent thinkers, and, in fact, it is still very much present in contemporary psychology." *An Introduction to Perception* (New York, 1975), p. 14.

(1713), to be essentially misguided and on the wrong track. At the very least, these works are not thought of as the kind of mainstream documents capable of bringing about a revolution in thinking but instead as expressing views that are exotic and out-of-the-ordinary. George Pitcher, for example, puts the prevailing assessment of Berkeley's ideas very well when he says that "Berkeley's metaphysics rises in the garden of British thought like some fantastic plant—beautiful and extravagant."[5] To the extent that the interest in Berkeley's *New Theory of Vision* is thought to lie in its contribution to Berkeley's metaphysics, it is treated not as remarkably successful but as remarkably unsuccessful.

What I want to do is to reverse this set of judgments. My primary endeavor is to find a way of reading the *New Theory* that accounts for its success as a contribution to the theory of vision. This enterprise requires a reorientation of the way Berkeley's project in the *New Theory* is frequently approached. Instead of reading the *New Theory* as a preliminary version of his later metaphysics, we should understand it as principally addressed to a positive program for solving some problems in the theory of vision. I intend to argue, moreover, that a richer understanding of the project of the *New Theory* will also shed new light on Berkeley's wider projects and show them to be less exotic and less misguided than has been supposed. It will be my contention that, when the *New Theory* is regarded as an example of a successful theory of vision, a somewhat different picture of its basic motivation and argumentation emerges than when it is read as a half-way house to Berkeley's metaphysics. I believe this new picture of Berkeley's motivation and argumentation is in fact a useful tool for interpreting Berkeley's views as found in *Principles* and *Three Dialogues*. I will be proposing that *Principles* and *Three Dialogues* be read in light of the doctrines found in the *New Theory* instead of trying to understand the *New Theory* through the lens of *Principles* and *Three Dialogues*. The *New Theory* can be read as a case history, illustrating, in a specific example, the more general claims that are made in *Principles* and *Three Dialogues*. On my account, the *New*

5. George Pitcher, *Berkeley* (London, 1977), p. 4. It should be noted that Pitcher does not describe the work of the *New Theory* as fantastic but instead as sober and well-informed.

Theory is a remarkably successful special case of a general theory that has not, on the whole, been thought to have been especially successful. What I hope to do is to remove the exotic and misguided flavor not just from the *New Theory* but from *Principles* and *Three Dialogues* as well. My suggestion is that—even though it is not possible to say of *Principles* and *Three Dialogues*, as it is of the *New Theory*, that they constitute a revolutionary moment—these two works can be shown to have been intended by Berkeley, like the *New Theory*, to address the mainstream theorizing of his day.[6]

In what follows, I concentrate my attention on developing an adequate account of Berkeley's revolution in vision as it is found in the *New Theory.* I nevertheless give a sketch of what I see as the implications of my interpretation of the *New Theory* for a way of reading *Principles* and *Three Dialogues*, although the actual reading must be deferred for another time. The present work falls into three parts. In the first part, I develop an account of Berkeley's motivation in writing the *New Theory* by situating Berkeley's project against comparable theories of his time. The second and largest part consists of a commentary on *The New Theory of Vision*. I have adopted the method of commentary because I have come to believe there is no better way of understanding Berkeley's arguments than to follow them step by step. Berkeley's arguments have often been admired even when his conclusions have been taken to be incredible. I have found, however, that, if one is careful to limit the conclusions Berkeley is thought to be drawing to what his argument can actually support, they are often less incredible than has been supposed. In the final part, I sketch an account of what I take to be the relevant connections between the *New Theory* and *Principles* and *Three Dialogues*.

My guiding aim has been to understand Berkeley's success. In this attempt, I have followed two methodological principles. I have tried, insofar as it remains possible, to identify the positions against which he is arguing by isolating the historical context of his arguments. I do not believe the best way to learn from a historical figure

6. C. M. Turbayne, in *The Myth of Metaphor*, rev. ed. (Columbia, S.C., 1971), also suggests that the linguistic model, whose success was demonstrated in the *New Theory*, can have a wider application against mechanism, although Turbayne does not work out this suggestion in any detail with respect to *Principles* and *Three Dialogues*.

is to take him to be motivated by the same questions and concerns that trouble a modern reader. Rather, I think the best way to learn something new is to be sensitive to what is novel from thinkers of the past, by respecting the extent to which their concerns are different from our own. Such sensitivity requires a full understanding of the context in which a particular line of inquiry emerges. I have also tried to give as sympathetic a reading as is possible. I have tried to give a reading under which Berkeley makes sense, is consistent, and is not prone to error. I am not claiming either Berkeley himself or my account of what he says is free from error. But I do not find attempts to refute or expose flaws in Berkeley's arguments, particularly when such attempts are based on the assumption that he must be wrong, a particularly helpful way of understanding what Berkeley is saying. The working assumption that Berkeley is making sense seems to me to be not only a more productive way of uncovering Berkeley's positions but also to increase the chances of discovering what is of genuine philosophic interest in Berkeley's concerns. In offering a sympathetic interpretation, I do not pretend to be giving a defense of Berkeley. I do not take it to be a part of my project to indicate whether or not Berkeley was right in what he says about vision. Such a ruling would require expertise far beyond what I can claim for myself, since it would be necessary to refer not just to Berkeley's texts and their historical antecedents but also to subsequent developments in the theory of vision.[7] My goal is more modest: to put forward what an adequate assessment ought to begin with, a thorough understanding of what is being claimed and why.

What Is the New Theory of Vision *All About?*

Difficulties in understanding Berkeley's *New Theory* begin with difficulties in understanding the nature of his project. Many accounts of what the *New Theory* is about run into problems that illustrate very clearly the dangers of reading the *New Theory* as directed to the same metaphysical aims that are often assumed to be

7. Robert Schwartz has done some work along these lines. See, for example, his "Seeing Distance from a Berkeleian Perspective" (unpublished).

governing *Principles* and *Three Dialogues*. Berkeley begins his *Essay towards a New Theory of Vision* with a paragraph in which he lays out his project:

> My design is to shew the manner wherein we perceive by sight the distance, magnitude, and situation of objects. Also to consider the difference there is betwixt the ideas of sight and touch, and whether there be any idea common to both senses. (*NTV* 1)

This passage is so straightforward and so bald a statement of purpose as to make any attempt at paraphrase seem ridiculous. Berkeley says he is going to show how it is we see distance, size, and situation. His project has an essentially positive thrust. Distance, size, and situation are all things we perceive by sight, and Berkeley is going to show how it is that we are able to accomplish these tasks. He also says he is going to look into the differences among ideas of sight and touch, with a particular eye to the question whether any ideas of sight are the same as ideas of touch. He is going to ask whether any of the things we learn by seeing are the same as what we learn through touch. There is at least an implication, although he does not say so, that an investigation of how we see distance, size, and situation will be peculiarly informative on the question whether the ideas we get by seeing are the same as or different from the ideas we get by touch. It is plausible that Berkeley is interested in investigating space perception because he thinks we will learn something interesting about the relation between the ideas of sight and of touch. The issue of the heterogeneity of ideas of sight and touch, then, would lie at the center of Berkeley's concerns. This impression is borne out by the structure of the argument of the *New Theory*. Each of the separate sections on distance, size, and situation end by drawing out the implications of that section with respect to the heterogeneity of the ideas of sight and of touch. Going by what Berkeley says in this first paragraph, then, he wrote *An Essay towards a New Theory of Vision* in order to show that a correct account of the way in which we determine by sight the distance, size, and situation of objects will reveal something about the relationship between ideas of sight and ideas of touch. His ultimate concern is to correct the errors of those who suppose we can learn the same things, get the same ideas, by sight and touch. Understanding Berkeley's motivation for writing the *Essay*, then, will require understanding why the issue of the heterogeneity of ideas of sight and touch is important to him.

Although there have been interpretations of the *New Theory* that take the heterogeneity issue to be central, most notably that of C. M. Turbayne, it is nevertheless surprising to find that much contemporary discussion of the *New Theory* highlights other issues, giving a somewhat different sense of Berkeley's project than the one he himself so starkly lays out.[8] For one thing, it is not at all uncommon for commentators to ignore the last two matters Berkeley raises, namely the visual perception of size and of situation, and instead to discuss the *New Theory* as if it were just about distance perception. D. M. Armstrong, for example, in his book about the *New Theory* pays almost exclusive attention to the discussion of distance, on the grounds that "the main themes of the *Essay* are all developed in it. The other three do little more than fill out and illustrate the argument."[9] In concentrating on distance to the exclusion of other issues, commentators have given Berkeley's project a predominantly negative cast. The main theme with respect to distance is to show, in George Pitcher's words, "that we are all perpetually the victims of an illusion when we think the things we see look to be various distances from us."[10] Berkeley's purpose in writing the *New Theory* is taken to

8. See C. M. Turbayne's "Editor's Commentary" to his edition of Berkeley's *Works on Vision* (Indianapolis, Ind., 1963); "Berkeley and Molyneux on Retinal Images," *Journal of the History of Ideas*, 16 (1955) 339–55; "The Influence of Berkeley's Science on His Metaphysics," *Philosophical and Phenomenological Research*, 16 (1956) 476–87; and *The Myth of Metaphor*. This last work is an extended examination of the implications of the approach to Berkeley that sees him as putting forward a positive account of vision, modeled on language, to replace a geometric theory. Louis Loeb also argues for the importance of the heterogeneity thesis, linking it, as does Turbayne, with Berkeley's claim that vision is a language. See Loeb, *From Descartes to Hume: Continental Metaphysics and the Development of Modern Philosophy* (Ithaca, N.Y., 1981). D. M. Armstrong takes the heterogeneity issue to be one of Berkeley's main concerns, but he interprets it quite differently, and he also attributes to Berkeley the goal of showing that "whatever is immediately seen has no existence outside the mind." *Berkeley's Theory of Vision* (Melbourne, 1960), p. xi.

9. Armstrong, *Berkeley's Theory of Vision*, p. 1.

10. Pitcher, *Berkeley*, p. 20. I do not mean to suggest that Pitcher limits his discussion of Berkeley's *New Theory* to distance perception. In fact, he gives very full discussions of all four sections of the *New Theory*. Pitcher's account of Berkeley's *motivation*, however, speaks only of Berkeley's need to reconcile his metaphysics with the apparent fact that things do look to be located at different distances in space.

be to show that, since distance is not immediately perceived, we do not, as we think, see objects at a distance from us; and hence objects, or visual objects at any rate, don't exist without the mind. Berkeley's motivation for writing the *New Theory,* then, is supposed to be directly connected to his idealism.

In support of this position, commentators typically make reference not to Berkeley's statement of purpose in the *New Theory* but instead to a passage in the *Principles* where Berkeley explains why he wrote the *New Theory.*[11] There he says:

> For that we should in truth see external space, and bodies actually existing in it, some nearer, others farther off, seems to carry with it some opposition to what hath been said, of their existing no where without the mind. The consideration of this difficulty it was, that gave birth to my *Essay towards a new Theory of Vision*, which was published not long since. (*PHK* 43)

Berkeley's target, based on this passage, is taken to be those who think we directly or immediately see distance and objects sitting at a distance, waiting to be seen. His target is our ordinary commonsense assumption that the things we see are sitting out there in space. His purpose is supposed to be an argument to show we are not really seeing things in space but only ideas of things "in the mind." Thus, on this reading, the *New Theory* is construed as a kind of half-way house or trial run at the full-blown idealism of the *Principles of Human Knowledge*.

This account of Berkeley's project which derives from the *Principles* does not, however, cohere very well either with Berkeley's statement of purpose in the *New Theory* or with the approach he follows there. For one thing, Berkeley doesn't say in the *New Theory* that he is going to show how we *don't* see distance or how seeing distance is illusory. Instead, he says that he is going to show how we *do* see distance (even though, presumably, distance is not immediately seen). There is nothing in the way Berkeley initially sets up his project in *NTV* 1 to suggest he thinks anything other than that we do make distance estimates by eye, that things do look to be at a distance

11. See, for example, Pitcher, *Berkeley*, p. 5; also G. J. Warnock, *Berkeley* (London, 1953), and I. C. Tipton, *Berkeley* (London, 1974).

from us. Although he goes on to indicate there are difficulties in understanding the nature of distance perception, it is hard to read Berkeley's first sentence except as saying that seeing distance is a phenomenon he is going to explain.

Second, there are difficulties in making out what Berkeley's argument on this reading is supposed to be. According to what we can call the "*Principles* account" of Berkeley's motivation, he is primarily interested in defending the claim that distance perception is "in the mind," and his demonstration hangs on the rather unfortunate assumption that whatever is "out there" cannot exist "in the mind" and whatever is "in the mind" cannot be "out there."[12] Thus Berkeley's demonstration that distance is in the mind is supposed to proceed by showing that distance is not immediately perceived, which is taken as tantamount to claiming that distance is not "out there." For this to be a plausible account of his procedure, Berkeley has to be assumed to imagine that if distance were immediately perceived, then distances would be "out there" and hence not be "in the mind." Berkeley, that is, must be taken to be equating an object's being immediately perceived with its being "out there."

It is certainly clear that by the time Berkeley wrote the *Principles*, he would not not have been moved by such assumptions. In *PHK* 42, he writes:

> Thirdly, it will be objected that we see things actually without or at a distance from us, and which consequently do not exist in the mind, it being absurd that those things which are seen at the distance of several miles, should be as near to us as our own thoughts. In answer to this, I desire it may be considered, that in a dream we do oft perceive things as existing at a great distance off, and yet for all that, those things are acknowledged to have their existence only in the mind.

It is obvious that Berkeley at this stage would not have thought that an argument that things just do look to be at a distance would be a reason for supposing the distance perceived is not "in the mind." It is very hard to imagine, moreover, why at any stage he might have thought that if something is immediately perceived then it is not in

12. See, for example, Pitcher, *Berkeley*, pp. 25–34.

the mind. Lights and colors are clear cases of things that undeniably are immediately perceived, but Berkeley nowhere seems to take our immediate perception of light and color as a reason for supposing that they are just there to be seen, that they have a clear extramental existence. Quite the contrary. Light and color are also prime examples of the sorts of things that exist "in the mind." So if, in fact, it were possible to make a case that distance, like light and color, is something we are built to register, something we "just see," this would have the effect of suggesting that distance perception, like color perception, would be dependent on the kind of sensory system we are built to have and hence would reflect this mind-dependence, would be "in the mind" and perhaps not "out there." It is very hard to believe, therefore, that Berkeley could have intended to equate a distinction between immediate and mediate perception with being "out there" as opposed to "in here," so that he could have supposed an argument that distance perception was not immediate could be an argument that distance was not "out there."

Finally, even if we assume that Berkeley had somehow tricked himself into a position where he was willing to equate being immediately perceived with being out there, it is still hard to find the argument he is said to be making in the text of the *New Theory*. According to the *Principles* account, Berkeley's problem is taken to be about our common-sense beliefs that things just do look to be at a distance because they just are at the distance they look to be. It is this common-sense set of beliefs Berkeley is supposed to be attacking that is equated with the claim that distance is immediately perceived. But whether or not this equation is correct, this is not a position Berkeley addresses in the *New Theory*. On the contrary, in the *New Theory* Berkeley does not even recognize this position as available, since he says that it is "agreed by all" that distance is not immediately perceived.[13] So we seem to have reached an anomalous position: Berkeley apparently rules out of court, on the grounds that no one holds it, the very belief he is later held to be saying was the one he wrote the *New Theory* to refute.

Part of the anomaly here stems from trying to read the *New Theory*

13. Turbayne, in his commentary to his edition of the *New Theory*, stresses that this claim is one that Berkeley adopts from his predecessors. *Works on Vision*, p. 19, note 10.

as if it were about our common-sense beliefs about sense perception. For the *New Theory* is not about common-sense beliefs at all. Berkeley is concerned to advance, as he says, a new *theory* to account for space perception, and his targets, the holders of the views he wishes to attack, are other theoreticians, in particular those who think space perception can be explained by reference to geometric reasoning on the part of perceivers. It is this group of theoreticians, and not ordinary perceivers, Berkeley has primarily in mind when he says it is "agreed by all" that distance is not immediately perceived. And the belief that he sets out to dispel in the *New Theory* is the far-from-common-sense belief that we perceive space by means of lines and angles. Difficulties in understanding the thrust of the *New Theory* have arisen from attempts to force an attack on common-sense beliefs onto an argument that is actually directed elsewhere. What seems to have happened is that the passage in the *Principles* has been given a "common-sense" interpretation. Berkeley has been read as setting out to refute a common-sense belief that distance perception is a matter of just seeing the things that are sitting out there. This common-sense interpretation is then imposed on the *New Theory* when the claim that distance is not immediately perceived is taken to express this same common-sense belief.

It is hard not to take Berkeley at face value when he talks in the *Principles* about the consideration that gave rise to the *New Theory*. But the way the passage from the *Principles* is usually read does not provide a motive that coheres very well with what Berkeley is actually doing in the *New Theory*. The discrepancies that have emerged between the *New Theory* and the *Principles* stem from the common-sense interpretation of the *Principles*. If, however, the project described in the *Principles*, of removing the belief "that we should in truth see external space and bodies actually existing in it," is not, in fact, limited to a common-sense belief about what we all think we see, then there will be considerably less reason for imposing this common-sense way of understanding what Berkeley is doing onto the more technical *New Theory*. One of Berkeley's notes to himself in the *Philosophical Commentaries* suggests that he does, in fact, link the problem raised in the *Principles*, that we think we see external space, with the beliefs of the optic writers considered in the *New Theory*. He says there:

> The common Errour of the Opticians, that we judge of Distance by Angles strengthens' men in their prejudice that they see things without and distant from their mind (*PC* 603).

In this note, Berkeley is claiming it is the supposition that we see geometrically which gives rise to the notion that we see external space. Berkeley's statement of his project in the *Principles* can be reconciled with his procedure in the *New Theory*, if we assume the consideration to which Berkeley is referring in the *Principles* passage is not just a problem about what exists in the mind. Berkeley is not asking, Do the things we see exist in the mind or out of it? Rather, he is concerned with theories of space perception which assume that what we see presupposes the existence of mind-independent, externally existing objects. Such theories might include common-sense theories, touched on in the *Principles*, that take what we see to be identical with external objects, but they also include the theories of the optic writers Berkeley mentions in the entry in his notebook, which form the target of the *New Theory*.

As the *Principles* passage is usually interpreted, Berkeley's focus is his idealism, his claim that what we immediately perceive are our own ideas. This focus has not altogether jibed with what is actually taking place in the *New Theory*, in part because the explicit target of the *New Theory* is geometric writers, who apparently share some of Berkeley's views about what is immediately perceived. Where Berkeley diverges from these geometricians is not in the account of visual experience that embodies his idealism but in his account of what our visual ideas represent. The geometric theory Berkeley discusses holds that our visual ideas represent mind-independent external objects, whereas Berkeley maintains they represent tangibilia. I am suggesting Berkeley's statement of purpose in the *New Theory* and the actual procedures he follows there make a good deal more sense if we see him as directing his attention to a theory of visual representation. He is making a case against those who think what we see represents bodies existing in external space. His account rests on a theory of space perception that makes no such assumption. On this approach to what Berkeley is doing, the problems he is trying to solve require a positive theory of how distance, size, and situation can be perceived by sight. They cannot be dissolved by a claim that we don't really see distance because distance is really "in the mind." I shall eventually be arguing that what is of primary importance to Berkeley's account is his theory of visual representation rather than what is often stressed, his theory of visual experience, which is to say, his idealism. It is the theory of visual representation defended in the *New Theory* that Berkeley later develops into an attack on materialism. Berkeley's immaterialism, I shall claim, should be distinguished from his idealism and identified with his theory of sensory representation.

Central to the approach to Berkeley for which I am arguing is the claim that Berkeley is putting forward an account of space perception because of problems he took to exist in what he set up as a rival account, the geometric theory. A complete understanding of Berkeley's motivation will therefore require starting from the theory of geometrical optics Berkeley is attacking. Before we can understand the nature of the problems that troubled Berkeley, we need an account of how it is that geometrical optics could have been taken by Berkeley to give rise to the belief about external space he regards as suspect. What is also needed is an understanding why Berkeley thinks an important issue bound up with his account of space perception is the heterogeneity of ideas of sight and touch. We turn to the first matter right away; it will take more time to get to the bottom of the second.

2

Geometrical Optics and Descartes

In the *New Theory*, Berkeley is very clear that the received view against which he is arguing is a geometric theory of perception, one held by what Berkeley refers to as "the geometricians." A good understanding of the nature of Berkeley's motivations, then, requires a picture of the role geometrical optics plays for him. In particular, the question that needs answering is why Berkeley thought a geometrical approach to space perception is peculiarly conducive to the belief that what we see are bodies existing in external space. An answer to this question, however, is a little more difficult to find than might at first be supposed, because it is not a straightforward matter to identify what Berkeley has in mind by "geometric optics." A number of different theories deserve the name "geometric optics," not all of which clearly constitute Berkeley's target. In the absence of a single monolithic theory, it is necessary to reconstruct which aspects of whose theories Berkeley might have had in mind. Such a reconstruction is of necessity speculative. There is no possibility of specifying with any certainty, in the absence of specific textual references, whose theory Berkeley is addressing.[1] It is possible, however, at the very least to rule out those

1. The *New Theory* is not, of course, completely without references to other authors. Berkeley added a footnote to *NTV* 4, a passage about the geometric account of how (near) distance is perceived, which reads: "See what Descartes and others have written on this subject." *NTV* 29 contains a long quotation from Isaac Barrow, *NTV* 30 mentions (citing Barrow) Tacquet's *Catoptrics*, and *NTV* 40 refers to Molyneux's

aspects of what can be called a geometric theory of vision in which he was clearly not interested. From such a winnowing a somewhat better picture of his target emerges.[2]

The literature with which it seems reasonable to suppose Berkeley was familiar used geometric optics in two different contexts. A treatise, such as that of William Molyneux, to which Berkeley frequently refers, consists almost entirely of demonstrations in applied geometry.[3] Molyneux's subject, however, is the physics of vision. He writes about the behavior of light rays in the presence of various sorts of lens. Molyneux specifically exempts himself from discussions of the functioning of the sensitive soul, from the problem of how perceivers see. A geometric optics of this sort is not intended to have psychological consequences. This is not the sort of project with which Berkeley had any quarrel. He tells us in the *Theory of Vision Vindicated* that his sole concern is to distinguish this geometrical enterprise from that of giving an account of how we see. It is this second problem that Berkeley takes to belong to philosophy, although today it would be relegated to psychology. He says:

> To explain how the mind or soul of man simply sees is one thing, and belongs to philosophy. To consider particles as moving in certain lines, rays of light as refracted or reflected, or crossing, or including angles, is quite another thing, and appertaineth to geometry. To account for the sense of vision by the mechanism of the eye is a third thing, which

---

Treatise of Dioptrics. All these references concern the Barrow problem. *NTV* 75 refers to "Gassendus, Descartes, Hobbes and several others," an article by Molyneux in *Philosophical Transactions*, and an article in the same journal by Wallis. All these references deal with the moon illusion. *NTV* 89 mentions Molyneux in connection with the problem of retinal inversion. *NTV* 125 cites Locke with respect to the issue of abstraction, and *NTV* 130 and 132 mention Locke and Locke and Molyneux on the Molyneux problem. In an appendix, Berkeley defends himself against a charge of attacking a strawman in his refutation of geometric optics by quoting Descartes, and he also quotes Gassendi on the moon illusion. It is significant, I believe, that all these references deal with specific problems or illusions except those which cite Descartes. Descartes is the only name Berkeley mentions with respect to what might be considered an overall geometrical theory of perception.

2. For another extended account of geometric optics and how it contributed to a geometric model, see Turbayne, *The Myth of Metaphor*, chaps. 6 and 7. Turbayne is a little more sanguine than I am, in talking about *the* geometric model against which Berkeley is arguing.

3. William Molyneux, *Dioptrica Nova* (London, 1792).

> appertaineth to anatomy and experiments. These two latter speculations are of use in practice, to assist the defects and remedy the distempers of sight, agreeably to the natural laws obtaining in this mundane system. But the former theory is that which makes us understand the true nature of vision, considered as a faculty of the soul. (*TVV* 43; see also *TVV* 37)

Berkeley's target, therefore, is only those theories which use geometry as part of a psychology of vision.

Unfortunately, it is still not possible to identify a straightforward psychological geometric theory. Berkeley's most likely sources for such a theory can be narrowed down to Descartes and Malebranche, both of whom used a geometric model not just to explain the physics of vision but also as an account of how perceivers see. It is natural to turn first to Descartes as a source for the beliefs Berkeley finds problematic, since Berkeley mentions Descartes by name, as one who holds that distance perception occurs "as if by a natural geometry."[4] Unfortunately, Descartes's theory, as it turns out, is not as clear as one might hope. It is possible to construct one sort of account of what Descartes's geometrical account of space perception is like, on the basis of passages from his Replies to the Sixth Set of Objections. But this account does not cohere as well as might be expected with his discussion of space perception in the *Dioptrics*. Descartes's theory, therefore, contains some unresolved tensions and cannot be used straightforwardly as a model for Berkeley's target. An attempt to resolve these tensions can be found in Malebranche's writings on space perception, and, although Malebranche is not mentioned by name in the *New Theory*, it is clear from Berkeley's *Philosophical Commentaries* that he had read Malebranche carefully.[5] It seems reasonable that the theory Berkeley called geometric optics, and probably thought of as held by Descartes as well as Malebranche, is most like the version found in Malebranche.[6] I want to take some

4. This citation occurs in the appendix added to the second edition in order to rebut the charge that Berkeley was addressing a strawman, and the passage cited is Descartes, *Dioptrics*, VI, 13.

5. Malebranche is mentioned by name in *PC* 230, 255, 257, 265, 269, 288, 358, 388, 424, 548, 686a, 800, and 818.

6. The relationship between Berkeley and Malebranche has been the subject of much discussion, and it has been argued, notably by A. A. Luce, *Berkeley and Male-*

time, however, to explore the ramifications and complications of the geometric theory as it appears in both Descartes and Malebranche. An awareness of the presence of these complications makes Berkeley's response to the geometric theory as he construes it, and various of the lines he takes, easier to understand.

The Sixth Set of Replies

In a relatively brief section, section 9, of his Sixth Set of Replies, Descartes makes a number of assertions that suggest a theory of space perception.[7] In particular, from this passage, it seems Descartes holds that observation embodies a piece of geometric reasoning such that, on the basis of what we see, we can conclude we know there is an external object, and we know its location in space, its size and shape. In this section Descartes distinguishes what he calls three grades of sensory response. His purpose in making this distinction is to be able to illustrate his claim that it is the intellect alone which can correct judgments about sensory events. The first grade of sensation is one we share with animals and is entirely physiological, consisting, Descartes says, "in nothing but the motion of the particles of the organs, and any change of shape and position resulting from this motion" (p. 294). The second grade, on the other hand, is mental and "comprises all the immediate effects produced in the mind as a result of its being united with a bodily organ which is affected in this way" (p. 294). Among the examples Descartes lists are "pain, pleasure, thirst, hunger, colours, sound, taste, smell, heat, cold and the like" (p. 294). Descartes appears to have in mind what are sometimes called secondary qualities; in this case, subjective states of the perceiver in response to some physiological change in the sensory system. The third grade Descartes describes as including "all the

---

branche: A Study in the Origins of Berkeley's Thought (Oxford, 1934), and recently by Loeb, *From Descartes to Hume*, that Berkeley ought to be seen as primarily a disciple of Malebranche and not, as I am suggesting, an opponent. For a thorough and balanced account of the relation Berkeley bears to Malebranche, see Charles J. McCracken, *Malebranche and British Philosophy* (Oxford, 1983).

7. *The Philosophical Writings of Descartes*, vol. 2, trans. John Cottingham, Robert Strothoff, and Dugald Murdoch (Cambridge, 1984), pp. 294–96.

judgements about things outside us which we have been accustomed to make from our earliest years—judgements which are occasioned by the movements of these bodily organs" (p. 295). Thus the third grade includes the beliefs Berkeley is particularly interested in, beliefs about *external* bodies. Inasmuch as both the second and the third grades of sensory response are part of our mental life, Descartes seems to be proposing a two-stage theory of perception, moving from the awareness of sensory states in us to beliefs about the properties of bodies outside us.

When Descartes gives an example illustrating the three grades, he specifically links the third grade, beliefs about external bodies, with the perception of geometrical properties. His example is a visual one, that of seeing a stick. The second grade, Descartes says, consists solely in the apprehension of light and color, and this is all that can truly be said to be sensory. He goes on to describe the third or intellectual grade as follows:

> But suppose that, as a result of being affected by this sensation of colour, I judge that a stick, located outside me, is coloured; and suppose that on the basis of the extension of the colour and its boundaries together with its position in relation to the parts of the brain, I make a rational calculation about the size, shape and distance of the stick: although such reasoning is commonly assigned to the senses (which is why I have here referred it to the third grade of sensory response), it is clear that it depends solely on the intellect. I demonstrated in the *Optics* how size, distance and shape can be perceived by reasoning alone, which works out any one feature from the other features. The only difference is that when we now make a judgement for the first time because of some new observation, then we attribute it to the intellect; but when from our earliest years we have made judgements, or even rational inferences, about the things which affect our senses, then, even though these judgements were made in exactly the same way as those we make now, we refer them to the senses. The reason for this is that we make the calculation and judgement at great speed because of habit, or rather we remember the judgements we have long made about similar objects; and so we do not distinguish these operations from simple sense-perception. (p. 295)

The upshot of this account seems to be that, in sense perception, I move by means of calculations from an awareness of the sensory state of light and color in me to a judgment about the distance, size, and shape—that is, the geometric qualities—of an object I judge to be external, "located outside me." These calculations, it should be noted, are psychologically real; at one time, at any rate, they were actually performed by the space perceiver. Although based on something occurring in me, they are not about something occurring in me. Distance, size, and shape judgments are not read off the sensory state but are instead the result of rational calculations. A result of Descartes's theory has been thought to be that, in vision, what we see is, thanks to an unnoticed intellectual judgment, perceived to be at the distance it is from us and to be the size and shape (the spatial extent) it actually is. The rational calculations by virtue of which we see distance have been taken to be about states of the external world. Thus it has been argued that Descartes's theory of geometrical optics constitutes an important underpinning of his general attempt to geometrize the study of nature.[8] The suggestion is that by means of his psychology of vision, Descartes is able to demonstrate that we have available to us an access to the primary qualities, or to the spatial location and extent, of the bodies surrounding us. This access, therefore, is said to guarantee the possibility of empirical confirmation of hypotheses.

This account of Descartes's geometrical optics and its implications hangs on a sharp distinction between the way we perceive what are often called secondary qualities like light and color and the way we perceive the so-called primary or spatial properties of distance, size, and shape. The perception of the former is sensory, and the perception of the latter is intellectual or judgmental. This distinction can then be assumed to support such claims as that the perceptions of secondary qualities are perceptions that don't resemble qualities of the external object, whereas the judgments about primary qualities do resemble qualities of the object. It is important to notice that this final claim, that the geometric qualities apprehended resemble qualities belonging to and which can be ascribed to external objects, the one that is all-important if Descartes's theory of vision is to provide empirical confirmation of geometric hypotheses, is an implication

8. See Nancy L. Maull, "Cartesian Optics and the Geometrization of Nature," in *Descartes: Philosophy, Mathematics and Physics*, ed. Stephen Gaukroger (Brighton, Sussex, 1980).

that is not explicitly drawn in the relatively brief passage in the Replies. On the basis of the Replies alone, there are certainly unanswered questions concerning the relationships between the judgments of spatial properties we make, the sensory states on the basis of which we make our judgments, and the actual disposition in space of external bodies that the judgments are about. In particular, we are given in the Replies no argument that would establish a distinction between that in vision which is sensed and that which is judged which might in turn establish the legitimacy or validity of our spatial judgments. Descartes refers in the Replies back to the *Dioptrics*, so that the implication is that the answers to these questions will be found in this latter work. In fact, however, the intellectual nature of the spatial properties of external bodies is not only not given explicit support in Descartes's lengthy account of vision in the *Dioptrics* but, even more surprisingly, appears to be contradicted there.

The Dioptrics

Descartes's purpose in writing the *Dioptrics* is somewhat different from that of the section just discussed from the Replies, since he is not, in the *Dioptrics*, particularly interested in looking at the peculiar role of the intellect. His stated goal in writing the *Dioptrics* was to give a sufficient grasp of the facts of vision to allow his readers to understand how new instruments, such as telescopes, function to improve or perfect our vision. Another goal that was perhaps at least as important to Descartes was to present an example of an entirely geometrically conceived science. If Descartes could show it is possible to give an account of vision that attributes nothing more than geometrical properties to the natural world, then he could use the success of such a science as a partial confirmation of the geometric picture of nature. Vision, or optics, is, moreover, a particularly important branch of science to choose, since it is the supposition that the qualities we see are like those of the natural world which might lead someone to reject Descartes's attempt to reduce the qualities of the natural world to extended matter in motion.

It is an important part of Descartes's project from the beginning to demonstrate the various ways in which qualities represented in us, as lights and colors in the case of vision, can be understood to represent qualities in the natural world by covarying with such natural qualities rather than by resembling them. Descartes, early on, introduces an analogy of a blind man with a stick in order to argue that, just as the blind man can become aware of vast qualitative differences in terrain just through pressures and impulses in his stick, which in no way resemble the qualities he is apprehending, so the colors we see can be understood to amount to the impact of light reflected off bodies on our eyes. Thus we will be prepared to agree that the ideas by means of which we represent are nothing like their causes, and hence, he says, the "mind will be delivered from all those small images flitting through the air, called *intentional species*, which worry the imagination of Philosophers so much."[9]

Given that the focus of Descartes's attention is on the ways in which the qualities we perceive fail to resemble qualities in the world they represent, it would not be surprising if he paid little attention to those perceived qualities which, it might be supposed, do resemble what they represent. What is striking about the *Dioptrics*, however, is that it is not just that Descartes doesn't bother to play up the differences between color perception and space perception but that, quite the contrary, his concern is to stress the similarities between the two. In the Sixth Discourse on Vision, Descartes claims the objects of sight can be reduced to six qualities—light, color, position, distance, size, and shape—of which only light and color are the proper objects of vision. This might seem to be a place where Descartes might make a distinction between sensory objects and objects of intellection, but he does not do so. His sole point here seems to be that light and color are the only qualities perceived *only* by sight, whereas the others are available to other sense modalities as well. Perception of light and color, he explains, are due to movements in those areas of the brain connected with the fibers of the optic nerves; light is the result of the forces of the movement, and the different colors can be traced to different "characters" of that movement. Thus Descartes's general picture takes the perception of light and color to be the natural, causal result of the way the optic system is constructed. If the par-

9. René Descartes, *Optics*, First Discourse, in *Discourse on Method, Optics, Geometry and Meteorology*, trans. Paul J. Olscamp (Indianapolis, Ind., 1965), p. 68.

ticular way in which I am aware is of red, then this is due to some characteristic sequence of physiological movements. A different sequence will result in my being aware of green and so on.

What is interesting is that this general picture doesn't change much when Descartes moves on to discuss space perception. "As to position," he says, "that is to say the direction in which each part of the object lies with respect to our body, we perceive this with our eyes in the same way as we would with our hands; and this knowledge does not depend on any image, nor on any action which proceeds from the object, but only on the position of the small points of the brain whence the nerves originate" (p. 104). When one of these positions is activated, then the perceiver is aware of the objects as lying in a particular direction. Descartes exploits the claim that direction is perceived by eye in the same way it is perceived by our hands to reintroduce the analogy of the blind man with the stick. This time, the blind man is determining the location of objects with two sticks that are crossed, and Descartes's point is that it is possible for the blind man to determine the location of objects along the length of the stick without needing to know where his hands are.[10] The moral for vision is that "similarly, when our eye or head turns in some particular direction, our mind is informed of this by the change which the nerves inserted in the muscles used for these movements cause in our brain" (p. 105). Descartes is telling a physiologically based story about events in the brain, which, covarying with direction, result in our being aware of a directional look to things. One muscle movement or another will result in something's looking to be in one direction rather than another. Our awareness of directionality is the result of the appropriate sequence of movements in the sensory system. There is no mention of a process of intellection of any sort and no reason to suppose that the representation of directionality is peculiarly suited to represent things as they are.

Nor does Descartes make any such claim when he moves on to discuss distance. He talks about several means by which we perceive distance, only one of which, and not the first, is geometrical in

10. Descartes is actually worried at this point about the problem of the inverted image on the retina, of which more later. See p. 26 below for a reproduction of Descartes's diagram of the blind man with the sticks.

nature, and he again begins the discussion by ruling out the use of images from the object. His primary goal is, as it was in the case of color perception, to replace an account in terms of intentional species, in which what we see is taken to resemble what is out there, with one in which what we see can be shown to be subjective responses to physiological changes in sensory states of the perceiver. He says:

> The seeing of distance depends no more than does the seeing of location upon any images emitted from objects; but in the first place upon the shape of the body of the eye. For as we have said, for us to see that which is close to our eyes, and to see what is farther away, this shape has to be slightly different. And as we change it in order to adjust the eye to the distance of objects, we also change a certain part of our brain, in a way that is established by nature to allow our mind to perceive that distance. And this we ordinarily do without reflecting upon it, just as when we squeeze some body with our hand, we adjust our hand to the size and shape of the body, and thus feel it by means of the hand without having to think of these movements. (pp. 105–6)

In this account of distance perception Descartes outlines the workings of a sensory mechanism, just as he did when talking about color perception. What he seems to have done is to have cast about for a physiological change, which covaries with distance, to which a subjective state of distance perception can be assumed to be a response. Our experience of changes in brain states, which are themselves changes brought about by changes in eye shape, results in a psychological state of things looking to be at some distance or other. When the eye is one shape, we say things look close to the eye, and when it is another shape, we say things look further away. But since 'looking to be one foot away' is simply a response to physiological movements, as 'looking red' is a response to a different set of physiological movements, there is nothing to suggest that the means that give rise to a 'one-foot-away look' more accurately reflect or resemble characteristics of objects in the external world than does a 'red look.'

Descartes does now go on to give a geometrical account of distance perception, although he does not single out this account in any way and, indeed, follows it by two further ways in which we come to perceive distance, which rely on pictorial cues. His geometric account of distance is as follows.

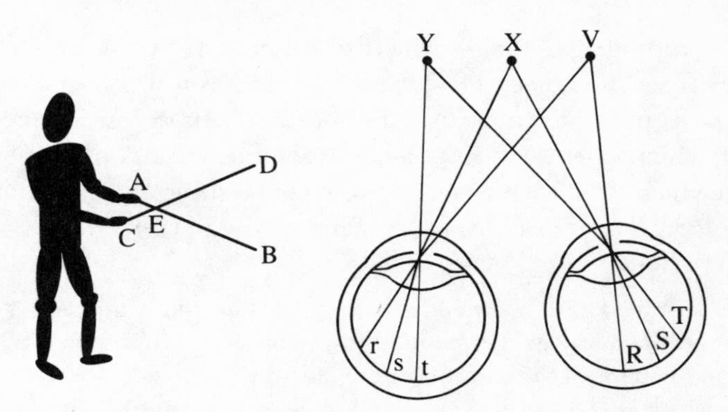

Figure 1. Descartes's comparison between the blind man with sticks and the geometry of the eye.

> In the second place, we know distance by the relation of the eyes to one another. For just as our blind man, holding the two sticks, AE, CE, of whose length I am assuming that he is ignorant, and knowing only the interval which is between his two hands A and C, and the size of the angles ACE, CAE, can from that, as if by a natural geometry, know the location of the point E; so also when our two eyes, RST and rst, are turned toward X, the length of the line Ss [the distance between the two eyes] and the size of the two angles XSs and XsS enable us to know the location of the point X. We can also do the same thing with the aid of one eye alone, by changing its position: as, if keeping it turned toward X, we place it first of all at the point S and immediately afterwards at point s, this will suffice to cause the magnitude of the line Ss and of the two angles XSs and XsS to combine together in our imagination, making us perceive the distance of the point X: and this happens by an action of thought which, although it is only a simple act of imagination, nevertheless implicitly contains a reasoning quite similar to that used by surveyors, when, by means of two different stations, they measure inaccessible places. (p. 106)

The first thing to notice about this account is that, unlike what Descartes says in the Sixth Set of Replies, it is not at all clear that he intends the "natural geometry" to indicate something psychologically real or, at any rate, to indicate an act of geometrical reasoning. He talks about "a simple act of the imagination" and describes the reasoning as "implicitly contained" instead of "very fast" or "habitual." So it would be in line with what Descartes says here to assume that looking to be at a distance is a response to changes in the optic angles and that the particular distances at which things look to be covary as if in response to the triangle formed from the optic angles. There is nothing in this passage to suggest Descartes thinks this particular way of perceiving distance is nonsensory or that the means by which distance is represented entitle us to conclude that the distance things look to be accurately reflects distances at which external objects are located. What seems to lie behind the geometric example is not an attempt by Descartes to show that distance perception can be seen as a piece of necessary reasoning. It seems more likely Descartes was encouraged to look to changes in optic angles to serve as distance cues through his desire to continue exploiting the analogy of the blind man with the stick. He is looking for something in the visual system that could be analogous to the angles at which the blind man's hands hold his sticks. Although, for the analogy to work, it is necessary for Descartes to assume we perceive the same sort of distance in the same sort of way by sight and touch, the motivation for the analogy was the refutation of the theory of intentional species, the theory that what we perceive resembles the object represented, and not a claim that seeks to locate some of what we think we see in the intellect.

We find Descartes primarily concerned in the *Dioptrics* with the constitution of the visual field, with the way things look to us. He is asking how it is that various visual ideas can get into us as perceivers. If I see a pen on my desk as black and as one foot away, Descartes is concerned to show that both of these ways in which I learn about the pen through vision can be understood as alterations in me responding to physiological changes, which can in turn be traced to an external, geometrically described object. Descartes is not very much concerned in the *Dioptrics* with beliefs we form about external objects as a result of information contained in the visual field. He mentions these beliefs only obliquely, when he is talking about illusions; when we look through a yellow glass without knowing it, we misascribe the color perceived to an object, and when we unwittingly look through a mirror, we similarly form a false belief about the location of the external object. Nor does he seem to be in any way concerned to argue that we are built in such a way as to be able, through our perceptions, to form accurate judgments about the distances at which external things are located. On the contrary, his point is that we are quite imperfect instruments for registering distance.

> It is also to be noted [he writes] that all the means that we have for knowing distance are very uncertain; for, as to the shape of the eye, there is no longer any noticeable variation when the object is more than four or five feet away from it, and even when the object is nearer, this shape varies so little that we cannot have any very precise cognizance of it; and as to the angles contained between the lines drawn from one eye to the other, and thence to the object, or from two positions of one and the same object, they also vary but little when what we look at is at even a short distance away from us. (p. 110)

Things do look to be at one distance or another from us, but this is not a claim that they look to be at the distances they are from us, since we are not very good at telling by sight how far away from us things are. The difficulty we have in judging distances by sight is very clearly illustrated when we consider very distant objects, like the sun and moon, which, Descartes says, have a look of being about one hundred or two hundred feet away, even though they are actually much further off.

None of the ways in which space perception seemed to be privileged in the Sixth Set of Replies receives much support in the *Dioptrics*. The distance things look to be cannot be ascribed to external objects directly, since it is very often the case that the distance at which things look to be is other than the distance at which they are.[11] Nor does the *Dioptrics* provide any support for the claim that distance judgments, unlike color judgments, are resembling—that our distance judgments involve a mental representation that resembles the distance about which we are judging. Such means of judging

11. Consider Descartes's concluding paragraph to the Sixth Discourse: "To sum up, in judging of distance by size, or shape, or color, or light, pictures in perspective sufficiently demonstrate to us how easy it is to be mistaken. For often because the things which are pictured there are smaller than we imagine that they should be, and because their outlines are less distinct, and their colors darker or more feeble, they appear to us to be farther away than they are in actuality" (*Optics*, p. 113).

distance as changes in the shape of the eye form too important a part of Descartes's theorizing for it to be plausible to suppose he is resting a great deal on a distinction between resembling and nonresembling perceptions.[12] Finally, the *Dioptrics* differs from the Sixth Set of Replies because it does not operate with a clear distinction between that which we perceive by the senses and that which, though ascribed to the senses, we actually perceive by means of the intellect. Indeed, the only distinction that seems to be present in the *Dioptrics* is between things perceived with only one sense and those perceived by more than one sense.[13]

12. I am therefore unable to accept Nancy Maull's claim that a passage from the Fourth Discourse of the *Dioptrics* shows that Descartes believed that "we may expect the same kind of resemblance between the object of our immediate awareness and the independently existing physical object as we find between an uncolored engraving and the reality it depicts" (Maull, "Cartesian Optics," p. 30). In the passage in question, Descartes is arguing against those who believe that perception takes place by means of little pictures in the head which resemble the objects seen. When Descartes talks about pictures in perspective, he is not actually talking about space perception at all. Rather he is saying, let's consider the actual case on which claims about little pictures in the head are modeled: the case of a picture that represents some scene. His point is that, even in this clear case of representation via resemblance, relatively little of what appears on the paper resembles the scene, and even in the case where we might expect to find some resemblance, namely, shape, the laws of perspective dictate that we depict a shape as, for example, a circle, by something that doesn't resemble it, as, an oval. He concludes by saying: "So that often, in order to be more perfect as images and to represent an object better, they must not resemble it. Now we must think in the same way about the images that are formed in our brain, and we must note that it is only a question of knowing how they can enable the mind to perceive all the diverse qualities of the objects to which they refer; not of [knowing] how the images themselves resemble their objects" (*Optics*, p. 90). Thus Descartes introduces the notion of pictures in perspective not as a part of a theory of space perception but to argue that, even in the best possible case, the resemblance between the representation and the object represented is highly imperfect. The overall thrust of the passage is to argue that, in perception, the representations by means of which we see need not resemble what they represent. For another discussion of Maull's argument, see Ronald Arbini, "Did Descartes Have a Philosophical Theory of Sense Perception?" *Journal of the History of Philosophy*, 21 (1983) 317–37.

13. Descartes gives an account of space perception in the *Treatise on Man*, ed. Thomas Steele Hall (Cambridge, Mass., 1972), which is similar to though less detailed than that of the *Dioptrics*. In the *Treatise on Man*, he also identifies spatial qualities as those which are "common to touch and vision, and even in some ways to the other senses" (p. 59) and gives an account of distance perception by triangulation which he again describes as taking place "as if through a natural geometry" (p. 62).

Geometry in Descartes's Theory

That Descartes refers back to the *Dioptrics* in the Sixth Set of Replies suggests that he takes the two accounts to be compatible. But, in fact, they are not about exactly the same things, and so using one to supplement the other is not a straightforward matter. The Sixth Set of Replies refers to intellectual judgments we make about the spatial properties of external objects. The *Dioptrics*, Descartes tells us, showed this is a matter for rational calculation, by showing that shape, size, and distance can be deduced from one another. Starting from what we see, we calculate that there is a body, external to us, of a determined shape and size and located at a determined distance from us. What Descartes calls the third grade of sensation is a rationally supported belief about the geometric properties of a physical object. There are several issues that remain puzzling or problematic on this relatively brief account. In the Sixth Set of Replies, Descartes stresses the intellectual origin of spatial information and asserts that the beliefs in question are about external objects. That perception is a two-stage process seems to mean, in this context, that it proceeds from the experience of sensible qualities to beliefs about the intelligible qualities of the external world. This has seemed to imply that Descartes perceived a connection between that which is apparent in sensation and the judgment that objects are located in external space, that geometric reasoning can move from the subjective state to the objective judgment. The nature of this connection is not entirely clear, however. Because I can deduce the size of an object from its shape and distance from me does not explain how I can calculate any of these properties from something else, namely, the sensation of light and color to which the second grade of sensation is limited, which does not have the size, shape, or distance of the physical object. It does not explain what justifies Descartes's apparent confidence in the visual system as an accurate instrument for locating objects in space.

The second issue that is a little difficult to get clear is Descartes's picture of the phenomenology of the visual situation. Even though the beliefs in question are intellectual in nature, we take them, Descartes says, to be the product of vision. We think we *see* an external object, and we think we are *seeing* its location, shape, and size. But these beliefs we hold, though true of an external object, are actually false with respect to what we see, since, for one thing, what we see is color, and external objects are not colored. So another problem is to understand how Descartes thinks a series of beliefs that are false about what we experience, but true of the physical world, become incorporated into and confused with what we experience. These puzzles arise because in the Sixth Set of Replies Descartes seems to be envisaging a situation in which the geometric qualities of the physical world become available to us in the course of our experiencing something else, namely, the way in which we are affected by that physical world.

It is not, as it turns out, possible to look to the *Dioptrics* for a more complete account of these matters, because the *Dioptrics* is not about the judgments we make about external objects; it is, instead, about the nature of our visual experience, about how we have been affected. Our visual experience is in the *Dioptrics* held to be not just of light and color but of light and color with certain spatial features. Part of the visual look is a shape, a size, and a distance. We experience the colored shape as being at one distance or another. As Descartes's discussion of the sun and moon shows, we can answer the question At what distance does the moon look to be? At what location are you experiencing it? without in any way suggesting that this is the distance at which an external object, the moon, is actually located. The moon is experienced as being not one or ten feet away but rather about one hundred feet away, and this is a fact that can have a psychological explanation. The questions Descartes addresses in the *Dioptrics* have to do with these phenomenally experienced spatial properties. He is asking how it is that our sensory system is able to be systematically affected by events in the external world so as to experience things as being, for example, at different distances. He wants to know what it is about perceivers that responds differentially to environmental changes so as to allow them to experience something as being some particular distance away. Descartes is talking, then, about the psychological representation of distance, an event in the life of a perceiver, that occurs when physical stimulation is fed through a sensory system. The distances, shapes, and sizes he is talking about here, though the result of objects of certain distances, shapes, and sizes, are to be understood as psychological states of awareness and not to be identified neatly with the distances, shapes, and sizes of these external objects. There is nothing in Descartes's account in this case, moreover, that would lead us to single out the spatial properties of the representation from other visual properties or allow us to assume they have a privileged status when it come to talking about the physical world. Descartes's theory of the perception of spatial properties does nothing to show why these spatial properties are the ones in terms of which we frame intellectual beliefs about external objects. It is true that in giving a causal account of psychological representations, mention is made of objects having spatial properties that serve as the initial occasions of stimulation, but the spatial properties of these causes are not a part of the visual experience itself and conclusions about these properties require independent justification.

Descartes has an account, on the one hand, of how we come to attribute spatial properties to external bodies, to be found in the Sixth Set of Replies, and an account, on the other, of how we come to have visual experiences of a spatial world, to be found in the *Dioptrics.* To the extent that these are considered accounts of two separate things, they are, of course, perfectly compatible, although the Sixth Set of Replies presupposes an account of space perception whose nature is not entirely clear. It is true that geometric calculations enter into both theories. Geometric calculations are central to the account of how we make judgments about the external world and also enter, rather less importantly, as one of the cues that account for distance perception. If these two occasions for geometric calculation are identified, either by Descartes or by his readers, then the compatibility of the resulting theory is less clear. What is perhaps more interesting is that we find the impact of Descartes's geometrical view of nature entering rather differently into the two theories. When we are talking about a sensory system, then the view that the natural world can be described geometrically encourages the construction of mechanical models. We account for what we sense by showing that it is the result of impact or impulses from one piece of matter to another. The result of carrying out such a mechanical account is to emphasize the lack of resemblance between the ways we represent, which are the result of ways in which we are affected, and the physical world represented. Thus we see in the

Dioptrics, where Descartes is talking about how we experience spatial properties, that the emphasis is on how imperfect we are as perceivers of spatial properties. When, as in the Sixth Set of Replies, Descartes is talking about how we make intellectual judgments about the spatial properties of external bodies, then he is dealing straightforwardly with geometrical judgments. In addition, moreover, geometry functions as a model for reasoning. Intellectual judgments are satisfactory to the extent they rely on necessary connections between the items in the judgment, and geometry is an example of such reasoning about spatial properties. The intellectual space about which we judge geometrically is not the same, however, and indeed does not apparently resemble the sensory space we experience. A geometric theory of space perception, therefore, can refer either to a theory that accounts for our beliefs about the spatial properties of the external world or to a theory that accounts for our ability to perceive spatially. When we are talking, then, about the geometric theory of space perception, it is going to be important to keep straight which aspect of geometry and what sort of space we are talking about. It is to Malebranche's account, which more explicitly distinguishes and makes use of these differing notions, that we must now turn.

3

Malebranche, Descartes, and Berkeley's Project

Malebranche's project has a great similarity to that of Descartes in that he, like Descartes, sees himself as demonstrating the viability of the intellect as a faculty of knowledge and as showing that we can be blind to its worth when we place excessive confidence in the senses. The overall thrust of Malebranche's work, especially in the *Search after Truth*, tends to be more negative than that of Descartes, however.[1] Malebranche is more interested in warning against the errors we are prone to than he is in following the results of relying on the intellect. His account of vision differs, then, from what we find in Descartes's *Dioptrics* in that Malebranche is less concerned with demonstrating the possibility of a positive theory of optics to replace one relying on intentional species than he is with exploiting a theory such as Descartes's in order to warn against errors that one can fall into through relying inappropriately on the senses. Our senses have been given us simply for the preservation and protection of our bodies, and we fall into error whenever we forget their dependence on our bodies and take them to be telling us about things existing independent of us. In developing this thesis,

1. Berkeley's effective knowledge of Malebranche's writings is thought to have been limited to the *Search after Truth*, which he probably read in the Taylor translation, *Father Malebranche his treatise concerning the search after truth*, trans. T. Taylor (London, 1700). See Genevieve Brykman, "Berkeley: Sa Lecture de Malebranche à travers le dictionnaire de Bayle," *Revue Internationale de Philosophie*, 114 (1975) 496–514.

Malebranche is frequently more explicit than Descartes in his handling of issues concerned with our ability to see the spatial properties of things in areas where information provided by the senses is most likely to seem to overlap with information from the intellect. Malebranche's greater specificity has the result of producing a theory that hangs together perhaps more clearly than that of Descartes, and it has the further result of playing up or dramatizing those features Berkeley took to be problematic.

The Sensation of Spatial Properties

Malebranche's theory of vision is given most fully in the first book of the *Search after Truth*.[2] His account of vision will reveal, he says,

> (1) that we should rely on the testimony of sight not in order to judge concerning the truth of things in themselves but only to discover the relation they have to the preservation of the body; (2) that our eyes generally deceive us in everything they represent to us: in the size of bodies, in their figure and motion, and in light and colors, which are the only things we see; (3) that all these things are not as they appear to us, that everyone errs regarding them, and that as a result we are plunged into an infinite number of other errors. (*ST*, bk. I, chap. 6, p. 25)

Malebranche's project is to demonstrate the misleading nature of *all* sensations. He begins his discussion with the geometric properties, extension, shape, size, motion, and distance, with the explicit purpose of showing that, insofar as our visual awareness of these properties shares the characteristics of sensation generally, we can be led into error. Visual sensations, according to Malebranche, are the expression of the union of mind and body. Seeing red is a way of being that a mind falls into when sense organs are affected by corporeal bodies. The laws of the universe laid down by God require there be no sensation in the absence of the appropriate physiological stimulation. What we see, then, is dependent upon retinal stimula-

2. Nicolas Malebranche, *The Search after Truth*, and *Elucidations of the Search after Truth*, trans. Thomas M. Lennon and Paul J. Olscamp (Columbus, Ohio, 1980).

tion, as well as the proper functioning of the rest of the optic system. An object cannot be seen unless it is projecting an image on the retina. Whatever knowledge, therefore, we might have about objects will not affect what we see unless it is somehow represented on the retina or in the visual system. For example, even though an astronomer knows the sun is much bigger than the moon, this knowledge is not represented in the visual system and hence is never a part of what is seen.[3]

Whatever is seen, what we learn with our eyes, is always relative to the sort of sense organ in operation and to the situation of the body perceived relative to that sense organ. *We* can never learn anything other than how a body strikes a perceiver of the sort *we* are. We see bodies as having a particular extension, but the extension we see is the result of the visual system we are using and hence will be very different from the extension seen by a mite; it will even differ, though less dramatically, from that seen by other humans. Some people, Malebranche points out, even see the same object as larger using the left eye than they do using the right.[4] When we recognize the dependence of the extension we see on the peculiarities of our own visual system, we will accept that we should never suppose we are seeing things as they are in themselves.

The nature of this argument requires Malebranche to play up the dependence of what we see on the visual system and so to emphasize the correlation between visual sensations and what is depicted on the retina. Nevertheless, it is often the case that what we see, the world as revealed by vision, is not the same as what is depicted on the retina. We see a cube, for example, as having equal sides, even though the image projected on the retina, a cube as seen in perspective, has sides that are unequal. In order to account for this phenomenon, that the way we see the cube is other than its image on the retina, Malebranche introduces the notion of natural judgments. What we see is informed by the judgment that the sides of the cube closer to us will project larger images than will those further away. We take account of the distance between the front and back sides of the cube in seeing it as having equal sides. Malebranche calls these

3. "Réponse à Régis," I, 4, p. 265, vol. XVII–I, *Pièces jointes et écrits divers*, ed. Pierre Costabel, Armand Cuvillier, and Andre Robinet (Paris, 1960).

4. *ST*, bk. I, ch. 6, sect. 1, p. 28.

judgments "natural" to emphasize that, although they occur in us, they are not made by us but, as he frequently says, often occur in spite of us. We are built so as to take account of distance, and hence seeing the cube as having equal sides is something that just happens to us. In speaking of natural judgments, Malebranche often emphasizes they are really a species of sensation or, as he sometimes says, compound sensations. Malebranche appeals to natural judgments whenever the visual world we see is other than what is recorded on the retina. Another example, very similar to the case just mentioned, is that of size constancy. When someone walks toward you, what is registered on the retina gets bigger, but what is seen is a person remaining the same size but getting closer. We are again said to be taking account of distance information when we perceive approaching objects as unchanging in size. These phenomena are like judgments because we account for the fact that what we see looks other than the way it is registered by referring to additional information in our possession or, rather, in the possession of the visual system. These ways we see are compound sensations because we are just built to respond to the information registered by seeing in the way we do.

Distance perception, for Malebranche, is similarly a matter of natural judgments. This means, although he does not spell this out, that Malebranche supposes that seeing distance involves more than what is registered on the retina.[5] The various ways we see distance that Malebranche describes are very similar to the account given by Descartes in the *Dioptrics*. Malebranche differs from Descartes chiefly in giving pride of place to the use of optic angles, but his account is otherwise similar. He says:

> The first, most universal, and sometimes the surest means we have of judging the distance of objects at a short distance is the angle made by the rays of our eyes with the object as its apex, that is, where the object is the point where these rays meet. When this angle is very great, we see

5. Or, rather, spells out rather circuitously the point that distance is not registered on the retina, by talking about motion. The diagram by means of which Malebranche makes his point about motion can be used equally well to make the claim that Berkeley says is "agreed by all" about distance. See the diagram reproduced on page 63 below.

> the object as very near; and when, on the other hand, it is very small, we see it as very remote. And the change that occurs in the state of our eyes according to the changes in this angle is the means the soul employs in order to judge the remoteness or proximity of objects. For just as a blind man could touch a given body with the ends of two straight sticks of unknown length and judge its approximate distance according to a kind of natural geometry by the position of his hands and the distance between them, so might the soul be said to judge the distance of an object by the disposition of its eyes, which varies with the angle by which it sees the object, that is, with the distance of the object. (*ST*, bk. I, chap. 9, sect. 3, p. 41)

The other means of judging distance Malebranche mentions are the feelings from the muscles as the eye is elongated or shortened, the size of the image on the retina, the distinctness or confusion of the image, and the possibility of taking account of intervening objects. Thus Malebranche, like Descartes, attributes to the visual system a variety of different sorts of information, both geometric and pictorial, and, like Descartes, he emphasizes that the geometric information, though more accurate, is most effective at short distances.

So, on Malebranche's account, to say humans can perceive distance by sight is to say that the visual system is built so as to register stimulation that correlates with distance, resulting in visual sensations of things looking to be at particular distances. Distance information is not available directly from the retina but requires various additional sources of stimulation for distance to be perceived. This means that we perceive distance by sight in the form of a compound sensation or natural judgment. Like any other sensation, the perception of distance is simply a way of being our mind falls into upon appropriate stimulation. Therefore Malebranche concludes his discussion of distance as he does that of other spatial properties, by pointing out that, as instruments for the registration of distance, we are not very accurate. We are built to be pretty good at assessing short distances by sight, but our assessments become much more rough-and-ready as distances get greater, until we really can't tell what distance things are at when they are very far away from us. We can't, for example, perceive by sight the distance the sun is from the moon; there is no distance the one looks to be from the other. Malebranche's moral is that since the visual system exists to help us preserve our lives, it is more trustworthy at short distances, where this information is more immediately relevant, than it is at great distances. Malebranche's discussion of space perception, then, appears to assume we can make claims about the existence of corporeal objects disposed about us in space. But his theory also assumes our perception of these objects is a construction of the visual system and as such is subject to error and illusion. For Malebranche, as for Descartes, the distances we see are representations of but cannot be identified with the distances at which things are located in space.

The notion of the natural judgment, therefore, allows Malebranche to make the claim that sensory information is entirely a matter of what the sensory systems are built to register while he also makes use of a difference between a "sensory core," which reflects retinal stimulation, and a "visual world," which we see.[6] By claiming that what is seen is often the product of a natural judgment, Malebranche is able to account for the fact that what we see is more informative than or different from what is registered on the retina without abandoning the claim that since sensation reflects only how we are built to process the impact of other bodies on our own, it cannot be used as a source of information about things in themselves. It is important to Malebranche that these judgments are merely "natural," they just happen to us, and they are not the result of our working out anything about how things are. In order to stress that natural judgments just happen Malebranche points out that they can subject us to illusions: a sun or moon on the horizon will look larger than when it has risen, even though the corporeal sun or moon has not altered in size, nor has the image on the retina.

Malebranche also emphasizes the complexity of the natural judgments we routinely make in sensation in order to demonstrate how unlikely it would be for such judgments to result from our own intelligence. In the Elucidation on Optics, Malebranche spells out this point:

> But to speak only about what concerns vision, God through this general law [that changes occurring in a certain part of the brain are accompanied by sensations of the soul] gives us precisely all those

6. The particular terminology I am using, a distinction between a sensory core and a visual world, does not appear in Malebranche. I take this terminology from Gary Hatfield and William Epstein, "The Sensory Core and the Medieval Foundations of Early Modern Perceptual Theory," *Isis*, 70 (1979) 363–84.

> perceptions we would give ourselves a) if we had an exact knowledge, not only of what takes place in our brain and in our eyes, but also of the situation and movement of our bodies, b) if in addition we knew optics and geometry perfectly, and c) if we could, on the basis of this actual knowledge and not of other knowledge we might have drawn from elsewhere, instantaneously produce an infinity of precise inferences, and at the same time act in ourselves according to these precise inferences and give ourselves all the different perceptions, whether confused or distinct, that we have of objects we see at a glance—perceptions of their size, figure, distance, motion or rest, and all their various colors.[7]

What we end up seeing, Malebranche claims, would be inexplicable unless we assume the presence in us of a knowledge of the laws of optics. We can account for the fact that, to use Malebranche's example, a child at ten feet away is seen as much smaller than a giant at thirty feet away, even though the images projected on the retina are the same size, only if we assume a knowledge of the distance between the child and the giant and a knowledge of the optical principle that images on the retina decrease in size at increasing distances. But since we ourselves do not possess such knowledge—indeed, what is required, Malebranche sometimes says, is a knowledge of divine optics—the right explanation is that God has given us the perceptions we *would* give ourselves *if* we had the requisite optical knowledge. When Malebranche appeals to a natural geometry in order to account for our perceptions of distance or size, he is emphatically *not* appealing to calculations that we, as opposed to our visual system, can be said to be making.

What we are seeing, then, is to be explained in terms of the operation in us, or rather in our visual system, of those geometrical and optical laws which link the situation and motion of our bodies with the situation and motion of other bodies toward which the visual system is directed. For this kind of geometric account to be a plausible explanation of what we are seeing, that we, for example, see the child and the giant as we do because God has equipped us

7. Elucidation on Optics, sect. 26, p. 733. Further on in this same Elucidation, at section 43, Malebranche illustrates this point in a long and fascinating passage in which he spells out all the natural judgments involved in seeing a large white horse at a distance of one hundred paces, running at full gallop to the right.

with optical and geometric principles, rests on certain assumptions about the child and the giant we are looking at.[8] Specifically, it assumes that the mind-independent child and giant are to be found in particular locations in space, where they take up a particular determined extent, or are of a determined shape and size. Without some assumptions about where and how big the object is that is being looked at, there would be no grounds for arguing that what we see can be explained only through such optical principles as that the retinal image decreases in size proportionally to increases in (actual) distance in (actual) space. A geometric account of perception like Malebranche's presupposes the truth of some geometric assertions about the object being looked at. Since the plausibility of the account rests on the assumption that the visual system is built to know its own causal history, the account therefore assumes the truth of a particular causal story, one that assigns geometric properties to the causes of what we see. Malebranche's account of sensation requires that we take what we see to be mind-dependent and hence not to be identified with the corporeal causes of what we see. His explanation for what is going on in the visual system assumes the possibility of our being able to make some claims about these corporeal causes, for example, that they have spatial locations. Malebranche's history of our mind-dependent sensations begins with geometrically described objects, disposed about us in locations in space, which act on our sense organs.

Natural Judgments and Free Judgments

Such geometric truths about the objects at which we are looking are not of course immediately available as a part of what we are seeing, since what we are seeing is not the object in space at which we are looking. The moral of the story about natural judgments is that their function is to construct for us a visual world, a way of seeing

8. I am here taking over Malebranche's terminological distinction between what we are seeing, a mind-dependent sensation, and what we are looking at, a mind-independent object. For a full account of Malebranche's somewhat varying use of this terminology, see John W. Yolton, *Perceptual Acquaintance from Descartes to Reid* (Minneapolis, 1984), chap. 2: "Malebranche on Perception and Knowledge."

things, out of the sensations we have, out of the various ways in which our body is affected. The visual world our sensory system constructs for us is one that will enable us to preserve and promote the well-being of our body. This visual world, however, is no more than a construct from the perspective of a being with our kind of sense organs from our location. Although the various natural judgments that go to make up this world are ones we cannot help but make, we should not, as we are habitually inclined to do, follow up these natural judgments with what Malebranche calls free judgments. We should not give free assent to the natural judgments, we should not believe that the actual world is the same as the one painted for us by our natural judgments. We should not freely believe, as our natural judgment inclines us to do, that snow is white, or that the moon is about one thousand steps away, or in general that the visual world we see is like the world that has an existence external to us.

Malebranche maintains we are prone to make this sort of habitual judgment in part as a result of another sort of natural judgment the visual system makes. In addition to the natural judgments that construct a visual world through geometric or optical principles, there is another sort of natural judgment, whose mechanism is not as clear, that invests this visual world with an "outness" or externality.[9] We unavoidably see the visual world as existing "out there," external to us. But since the visual world we see is a construct that is dependent upon our sensory system, the judgment that it exists out there, external to and independent of us, though unavoidable, is false. Malebranche explains this in a difficult and highly controversial passage from Book I of the *Search after Truth*:

> It seems to me beyond question that our souls do not occupy a space as vast as that between us and the fixed stars, even if it be agreed our souls are extended; thus, it is unreasonable to think that our souls are in the heavens when they see the stars there. It is not even thinkable that they should be projected a thousand feet from their bodies in order to see houses at that distance. Our soul, then, must see stars and houses where they are not, since the soul does not leave the body where it is

9. "Outness" is the word used by Malebranche's first English translator, Thomas Taylor, for this concept of externality.

> located, and yet sees them outside it. Now given that the stars immediately joined to the soul (which are the only stars it can see) are not in the heavens, it follows that everyone who sees the stars in the heavens and who then voluntarily judges they are there performs two false judgments, one of which is natural, the other free. One is a judgment of the sense or a compound sensation, which is within us, occurs independently of us, and even in spite of us, and according to which no judgment should be made. The other is a free judgment of the will, which can be avoided, and which consequently we must not make if we wish to avoid error. (*ST*, bk. I, chap. 14, sect. 1, pp. 67–68)

At the very least, Malebranche is stressing that what we see must be understood to be the result of action upon our sense organs and hence other than what it is that is actually sitting far away in space.

In his account of the false judgment to be avoided, Malebranche introduces a distinction between that which I perceive immediately and that which is known by means of what I immediately perceive. When I say I see the sun, there is something that I am aware of seeing, perhaps a small yellow disk. I see this small yellow disk as having an outness, as existing independently of and external to myself. This is a natural judgment, and it is false because what I am seeing does not exist independently of myself. I am led to make this judgment because I know that what I am seeing, my sensations, are independent of my will, and so, Malebranche says, "the soul is led to judge that they are external to it, and in the cause that represents them to it" (*ST*, bk. I, chap. 14, sect 1, p. 68). The sun I am seeing is not the same as the corporeal sun that causes what I am seeing and is what I am looking at, because the corporeal sun *is* independent of and external to me.[10] *Seeing* the sun requires an interaction involving the state of my sense organs and the state of my mind. So I am judging falsely when I invest my mind-dependent sensations of something small and yellow with some of the properties of its nonsmall and nonyellow cause, which is to say, when I judge that the small yellow disk is external to me and when I judge that it is in the heavens. These are actually properties not of the small yellow disk that I immediately see but of what I know by means of what I am immediately seeing.

10. Malebranche is using causal language here, although he will of course eventually require translations into his occasionalistic version of causation.

It is important to see that the externality or outness Malebranche is talking about here is a somewhat different concept from that of being at a distance, although there are certainly ways in which the two concepts are related. When Malebranche is talking about the natural judgment I make that what I see exists out there (*au dehors*) or outside of me, he is talking about the conviction I have that what I see exists independent of me, in the way that the cause of what I see exists, as it does, independent of my perception of it. I think the way the sun actually is is the same as the way I see it to be. But although this is a useful mistake from the point of view of my preservation, it is nevertheless a mistake, since the way I see the sun is dependent on my sense organs and my perspective. Malebranche also says I take the things I see to be "in the cause that represents them." It is at this point that distance judgments seem to be involved. What I do is to take what I am seeing, which I have confused with its cause when I suppose it to be mind-independent, to be located in that mind-independent spot in space where the cause is to be found. I identify the distance that is constructed in my visual field, which is where I see the sun to be, with the location in space where the cause of the distance perception is located. I invest my distance perception, that is to say, with an outness. Malebranche is not making the mistake he has been accused of making, that of supposing whatever is in the mind can not be at a distance and whatever is at a distance can not be in the mind. He is still assuming, however, that he has at his disposal, as a part of his explanatory arsenal, facts about the spatial location of the mind-independent causes of what we see.

Malebranche's point when he talks about the ways in which I confuse what I am immediately seeing with what I know by these means is again to emphasize the falseness of our natural judgments. For, as he says, when I am judging that the sun I see is external to me,

> Yet it can happen that we see this first sun, which is intimately joined to our soul, without the other being on the horizon, and even without it existing at all. Likewise we can see this first sun as larger when the other is rising than when it has risen well above the horizon; and although it is true that this first sun that we see immediately is larger when the other is rising, it does not follow that this other that we are looking at, or, rather, toward which we are turning our eyes, is larger.

> For it is really not the rising sun that we see or are looking at, since this one is several million leagues away. Rather, we see the former, which really is larger and such as we see it, because all things that we see immediately are always such as we see them, and we err only because we judge that what we see immediately is found in the external objects that are the cause of what we see. (*ST*, bk. I, chap. 14, sect. 2, p. 69)

These arguments, however, intended to convince us that there is a difference between what we see immediately and what we know about on the basis of what we see, go through only if we are entitled to assumptions about what we know about on the basis of what we see. Malebranche's explanation for the natural judgment by means of which we come to see the visual world as external and at a distance assumes we can make assertions about a causal world that is, unlike the visual world, external and at a distance from the perceiver. Similarly, his claim about the falsity of our natural judgments of the size and distance of the sun we see assumes the truth of other judgments about the size and distance of the sun toward which we have turned our eyes. It is only such assumptions that permit the explanation that the visual system is attributing to sensations the size, distance, and externality that are actually true of their causes.

Sensations and Ideas

The chapter on natural judgments of externality exploits a distinction between that which the soul sees immediately and what it knows by these means in order to emphasize the falsity of the natural judgments based on the former. What we are clearly owed is some account of the latter process that will presumably explain how it is that, by means of false sensations, we are able to know some true things about their causes. At first glance, the problem in Malebranche's terms seems insoluble, since he has also said it is possible to have a visual sensation of the sun without there being any corporeal sun existing to cause the sensation. But Malebranche repeatedly warns throughout this chapter that what he is saying will not be entirely understandable in the absence of his theory of ideas. So it is Malebranche's theory of ideas that is apparently intended to provide a bridge for what looks like the unbridgeable gap between the sensations that are dependent on the nature of the perceiver and their independent causes.

When Malebranche says we could have a sensation of the sun even if the sun did not exist, his point is that sensations depend for their content on the particular sense organs of the individual perceiver. We cannot account for what we sense simply by talking about the corporeal world, for we have to talk about our various capacities for sensation, the various means by which we apprehend the world, sensorily speaking. The upshot, as far as sensation is concerned, is that whatever glimpses we get of the world are always through our own sense organs and are from our own perspective. Sensations, Malebranche concludes, are states of being of the perceiver and hence not to be taken as adequate representations of the external world. Ideas, however, differ from sensations, according to Malebranche, precisely in not being dependent on the individual sense organs of the perceiver. Ideas are the product of our representational capacities (to be exact, the intellect) which go beyond the individual perspective of the perceiver. Ideas have a content that is not dependent on any individual mind and, for that reason, are capable of representing that which is independent of mind. It is because we have ideas in addition to sensation that we have knowledge that goes beyond our immediate sense presentations. The problem, then, is to see how Malebranche can identify this representational capacity belonging to ideas.

In a passage from the Tenth Elucidation, Malebranche spells out the different functions of sensations and ideas. He says:

> It seems to me worthwhile to point out that the mind knows objects in only two ways: through illumination [*par lumière*] and through sensation. It sees things through *illumination* when it has a *clear idea* of them, and when by consulting this idea it can discover all the properties of which these things are capable. It sees things through *sensation* when it finds no clear idea of these things in itself to be consulted, when it is thus unable to discover their properties clearly, and when it knows them only through a confused sensation, without illumination and without evidence. Through illumination and through a clear idea, the mind sees numbers, extension, and the essences of things. Through a confused idea or through sensation, it judges about the existence of creatures and knows its own existence. (p. 621)

When we have an idea, we have grasped an essence. On the basis of the content of the idea, of extension, say, we can discover all of the properties of extended things. Thus the contents of ideas are general and do not depend, in the way sensations do, on particular episodes involving some body. Since the idea is not limited to some set of transactions taking place in a body but instead has as its content all of the infinite possibilities of which extension is capable, it represents not extension-as-perceived-by-me but extension-in-general. The content of such an idea is independent of any particular perceiver and represents not the state of being of some perceiver but an essence independent of any perceiver.[11]

It is important to Malebranche's account that we, in fact, have ideas in the course of having sensations. "When we perceive something sensible," he says, "two things are found in our perception: *sensation* and pure *idea*" (*ST*, bk. III, pt. II, chap. 6, p. 234). The perception I have of a sensible object involves two different kinds of conscious mental contents. When I perceive a sensible object, I have a confused sensation of things like its color. I have this sensation only by virtue of a relationship my soul enters into with that other body and my own, according to the laws governing the union of soul and body, as Malebranche frequently says. It is only by virtue of such relationships that bodies are seen as colored. I learn only about the particular color I am sensing through that particular sensory transaction. Having a sensation of yellow, for example, won't tell me that I am the kind of being that can also have a sensation of red. It is because of the limited nature of the information I get from sensation that Malebranche calls sensations confused. The sensation, however, is accompanied by an idea of extension, since the color appears spread out as if on a surface. Malebranche says the idea of extension of the object is made visible by the color. The content of the sensation is entirely perceiver-dependent, but the nature of the content of the idea is not. I can represent what I see not only as having the particular spatial extent I see but also as having any other possible spatial extent. My representational capacities with respect to extension when I have an idea of something extended are quite different from my capacities with respect to color. This is why the

11. Notoriously, of course, it is Malebranche's contention that the intelligible extension we perceive is in God.

idea I have of extension can be identified as a general idea, one that does not depend, in the way my sensations do, on particular episodes involving my own body. Since the idea of extension I have is general, this generality would suggest I have the same idea, by means of which I can come to know any of the consequences of that idea of extension, no matter what sense modality is in operation, for the general idea I have is independent of any sense modality.

Thus, a particular sensory presentation informs us some object is present. Malebranche holds we cannot have an awareness of a particular thing except by having it strike our sense organs so that we are aware of it as from a particular point of view. The extension that is by these means made visible is particularized. So, as Malebranche has stressed throughout Book I, as particularized, any judgment about it may be false. But I grasp any particular extension, any particular shape, for example, only insofar as it is one possible shape among the infinite varieties that are equally possible. It is because of the clear idea I have of extension-in-general that I am able to grasp this particular way of being extended. The presentation I am having depends upon the idea of extension in general. If I assent to the existence of this-extension-in-particular, I may well be making a false judgment. On the other hand, I can discover all of those properties which are necessarily connected with this extension. Assertions I make about actuality are about what is given in sensation and may well be false. Assertions I make about possibility are derived from a knowledge of essences, and here I am capable of true judgments.

It is therefore Malebranche's insistence that perceptions involve both sensations and ideas which allows him to provide some sort of ground for claims he makes, as in his optics, that seem to require a knowledge of events in the corporeal world. The kind of knowledge that we have about the corporeal world is a knowledge of its essence.[12] We know the corporeal world is essentially extended, and so we know any account of the doings of the corporeal world will be

12. Or rather, this is what we know about the corporeal world when we know it by way of idea. When the corporeal world acts on our sense organs, then we apprehend it by way of sensation. In and of itself, that is, without interacting in any way with our faculties of apprehension, the corporeal world is invisible. Strictly speaking, since the ideas we have are in God, they can be said to be what God knows about the corporeal world.

geometric in nature. We know an account of vision will be at least an account of the action of extended bodies, disposed at particular locations in space, on the bodies that constitute our sense organs. Malebranche can argue that since particular sensory experiences embody ideas, they provide us with information, in the form of the idea of intelligible extension, from which we can discover any of the rest of those properties which will allow us to ground such geometrically based claims. Since the sensation we are having embodies one possible consequence made actual of the idea of intelligible extension, we are able to work back from it, via geometrical principles, to establish the optical laws of which this particular sensation is a consequence. This sort of knowledge, although it is knowledge about how perceivers work, is not knowledge we get *in perception*. Rather, it is intellectual knowledge that is part of the theory of perception.

Malebranche's account in Book I of what we see stressed repeatedly that what we see is an imperfect representation of the corporeal world. In making this claim he appeared to be helping himself to various claims about the corporeal world, as that the corporeal world consists of extended matter located in space, whose various shapes, sizes, and distances from the perceiver sometimes were like but often were not like the way their shapes, sizes, and distances were represented by the visual system of the perceiver. It turns out, however, that the mind-dependent sensations described in Book I embody ideas. Even though the particular perception makes a particularized extension visible, we are nevertheless capable of discovering all those properties necessarily connected with this presentation. We thus have at our disposal the means for working out the various spatial relations that must exist between what is seen, what is on the retina, and what exists external to the perceiver. We are capable of becoming theoreticians of perception. These various spatial relations, therefore, though not what we immediately perceive, can be said to be known on the basis of what we immediately perceive.

Descartes and Malebranche

The chief problem that emerged for an understanding of Descartes's theory of space perception is that of integrating his account in the Sixth Set of Replies, concerning the use of the intellect in the discovery of the spatial properties of the external world, with his account in the *Dioptrics* describing the visual system and how we come to see a spatial world. If these two accounts are run together, then it seems Descartes supposes the visual system is built so as to be capable of just seeing shapes, sizes, and distances as they exist in the external world. Such a supposition, however, is contradicted by at least some of what Descartes says in the *Dioptrics*, where he seems to think of the visual system as inadequate to represent things as they are. Malebranche's account differs from that of Descartes in being more explicit about the role it assigns to the intellect and to the various sensory systems; it thus provides an unambiguously integrated account.

The most difficult problem to resolve in the Cartesian account has to do with the status of the spatial information available to us through sight. In Malebranche's account, it is clear he is treating the problem of how it is we are able to *see* the distances, sizes, or shapes of things as a problem about the operation of the visual system, that is, as a problem about sensation. We see with our eyes, or with our sense organs for vision, and so what we see is always dependent upon the ways these organs work to register information. Malebranche's account follows the implications of Descartes's account of space perception as it occurs in the *Dioptrics*. Malebranche is quite explicit, however, that the ways in which we come to represent spatial properties cannot be identified with the spatial properties of the external world. Malebranche's account of how we come to construct a visual world with things at different distances and of various shapes and sizes describes the construction of the visual world as a three-step process. There is an initial sensory state in which the sense organs are activated from the corporeal world, for example, images appear on the retina, followed by two kinds of natural judgments. The first kind of natural judgment relies on geometrical and optical information in the possession of the visual system to produce a visual world in which things look to be at distances and to be of a shape or size. The second kind of natural judgment invests the size or shape or distance with an 'outness' or externality, through a misidentification of what is seen with its corporeal cause. It is clear from what Malebranche says that the natural judgments, and hence the natural geometry by means of which spatial information is registered, do not describe calculations perceivers have consciously performed. We know they take place only because, unless we attribute such calculations to the visual system, we have no explanation of what we end up seeing. This is the position it seemed Descartes was taking in the *Dioptrics* about natural geometry, although it appears to be contradicted by the Sixth Set of Replies. Malebranche also spells out that if we give free assent to these natural judgments, we will put ourselves in the position of believing false things about the external world. The world that is actually out there does not have the sizes or shapes we see them as having, nor are things necessarily at the distances they look to be. For Malebranche, then, when we ask How is it that we are able to see things as being at a distance or as being of a certain size? we are not asking how is it we are able to perceive the distances, sizes, or shapes things actually are in the external world because, in fact, how things actually are in the external world is never what we see.

The second problem that emerges from a reading of Descartes is that of integrating the spatial information available from the intellect, on which a geometric account of the corporeal world is based, with the spaces and spatial properties that we see. For if the spatial properties we see are dependent upon our sense organs and hence systematically misleading, then it is difficult to understand why spatial properties are more likely than any other properties to represent the actual nature of the corporeal world. How is it possible to claim that spatial properties are the province of the intellect? On Malebranche's account, the spatial properties made available to us in sensation are nevertheless intelligible, not sensible qualities. When I see color, the color appears to take up a particular spatial extent, but Malebranche describes this as a process in which the idea of intelligible extension is made visible. It would be quite wrong to think of the color as actually taking up space, since appearing red is just a state of being my mind falls into under appropriate stimulation and, as such, cannot take up any space. (Hence the particular amount of space the color appears to take up does not reflect any amount of space in the corporeal world.) The idea of intelligible extension is, however, a part of the experience insofar as I grasp the color as being of some shape or size. That this spatial information is intelligible as opposed to sensory information is established by thinking about our intellectual capacities with respect to this information, by thinking about what we can do with it. If I am having an experience of red, then this tells me nothing more than that I am a being who can have an experience of red, but from an experience of extension I can discover any property necessarily connected with it.

Malebranche therefore has a criterion that enables him to distinguish sensory information, information dependent upon particular sense organs, from intelligible information. Intelligible information is effectively infinite. What I know when I perceive something of a particular shape is not limited to the occurrence of that shape but includes all of the properties necessarily connected with it; when I perceive a particular color, however, I am limited to an awareness of the occurrence of that color. For this reason, the apprehension of sensible qualities can be seen to be dependent on the sensory apparatus of the perceiver, but the apprehension of intelligible qualities is not limited to the occurrence of some particular episode. Because what we know when we apprehend intelligible qualities is that from which all other properties can be discovered, the ideas of intelligible qualities are independent of any particular mind. Ideas of mind-independent intelligible qualities can therefore be used to represent an independent reality. Particular sensory experiences contain information that, if used appropriately, is effective for representing the essence of the corporeal world. We can use this particular information as a basis for calculations that will be descriptive of spatial properties of the external world. Thus Malebranche agrees with Descartes's position in the Sixth Set of Replies that the intellect makes available spatial information and that calculations performed by us will inform us about the external world. He also thinks this intelligible information is available in episodes that include sensation. But it requires explicit intellectual work beyond sensation to convert information included as a part of a sensory episode into judgments about the external world. We cannot learn about the actual distances, sizes, or shapes of external things simply by looking about us.

Berkeley's Project

In both Descartes's *Dioptrics* and Malebranche's *Search after Truth*, Berkeley could have found the problem of the perception by sight of spatial properties discussed as a problem in the psychology of vision. Both Descartes and Malebranche explain how it is that things look to be at different distances and of different shapes and sizes. Part of the terms of the problem, then, is that things do look to be at distances and of sizes and shapes. Since this is a problem in the psychology of perception, however, its subject matter is perceivers. The task is to say what it is about perceivers that enables them to see the different distances, shapes, and sizes of things. Part of the terms of the solution will be the ideas perceivers have about distances, shapes, and sizes. What is being explained is how it comes about things have a distance or a size or a shape look. Berkeley conceives his project to be a contribution to this same discussion, for the theories he takes to be rivals to his try to account for space perception, as Descartes and Malebranche do, in terms of a natural geometry. Thus Berkeley, too, is concerned with accounting for a particular phenomenon: the perception by sight of certain spatial properties, or how spatial information gets into or gets represented by perceivers. The problem of how we see, for example, distance is not a problem of whether or not there are objects at a distance. Indeed, Malebranche is explicit about the irrelevance for problems in the psychology of vision of issues about the existence of objects in the corporeal world. On the other hand, the terms of the problem certainly do require that distances, shapes, and sizes are among the things we perceive by sight. Inasmuch as the way we perceive distance, shapes, and sizes by sight forms the subject matter of the inquiry, there is no room within this way of formulating problems about distance perception for a claim that we don't really see distance or that distance perception is nothing but an illusion. So there is no reason, at least initially, to take Berkeley's excursion into the psychology of vision to be anything but an attempt to show how it is that we *do* see things like distance.

Both Descartes and Malebranche, it is true, stress that the way in which the visual system represents sizes or distances is not immediately reflective of the sizes or distances of external objects. There is a sense in which *these* theories have as a consequence that we are victims of a distance illusion. But these theories don't hold simply that we must be victims of a distance illusion because what we perceive is in the mind and being at a distance is not. Both Descartes and Malebranche are concerned to make the point that when we understand the nature of sensory operations such as seeing, which depend upon a sequence of organic changes in the sense organs, then we will understand the extent to which what we see depends upon the nature and perspective of the perceiver. Therefore, Malebranche argues, when we take what we see to be external to us, we are making a natural judgment that is false, because what we see is not external to us but dependent on our sense organs. Malebranche is distinguishing distance judgments from 'outness' or externality judgments. Both distance and outness are ways in which things look. It is always false, however, that the ways in which things look are mind-independent or external to the perceiver, whereas, thanks to the nature of our perceptual apparatus, things sometimes do and sometimes don't look to be at the distances they are. What makes all of these judgments possible is the conviction that since what we see is dependent on our sense organs, it cannot automatically serve as a representation of a mind-independent reality. These theories are committed to a distinction between a visual world that is the result of the operations of our sense organs and an external corporeal world. This is the distinction Berkeley is determined to break down. The claim that we are victims of a distance illusion is in actuality a consequence of the theories of Descartes and Malebranche that Berkeley is going to argue against.

Berkeley presents his project as opposed to the theories we have been discussing to the extent these theories rely on geometric principles. The sorts of difficulties (he holds) that rival theories fall into and that his own avoids stem from a misapplication of geometric principles to the psychology of vision. It has also seemed to be the case that Berkeley credits this same misuse of geometry with encouraging the problematic belief in an external world. To the extent the problem Berkeley shares with the other theories is that of explaining how we see, then the geometric reasoning he is concerned with is the natural geometry or natural judgments that are assumed to describe the way the visual system works and not the sort of reasoning the intellect has been held to engage in when it theorizes about the geometric properties of the external world. These are not thought to be judgments we voluntarily or knowingly engage in, and since they are judgments explaining what we see, they are not about the external world at all. Nevertheless, in the hands of Malebranche, it is clear such judgments do presuppose the presence of information about external objects. Malebranche maintains that when I take the object I see—as a result, let's say, of a natural judgment of triangulation—to be at a distance and also to be external to me, then what I am doing is seeing the object to be at the spot where the cause is located. I do not of course immediately see the spot where the cause is to be found. Malebranche seems to be relying on the fact that when I see distance by means of a natural judgment of triangulation, then the visual system has information about the location in space of the cause of what I see, even though this is not what I see. I, the perceiver, do not have this information, since I am limited to the effect in me of the operation of my visual system, this being what I see. We speak of natural judgments when what I see presupposes information not on the retina but assumed to be in the possession of the visual system, and among this information can be the location of an object causing what I see.

Malebranche's account of what we must attribute to the visual system, namely, calculations of lines and angles, does indeed presuppose that the lines and angles are effective means for seeing distance, because they are calculations involving the external locations of objects. Malebranche's calculations concern the spatial properties indifferently of perceptual objects and external or mind-independent objects. His account of how we see distance depends upon the supposition that what we see can be connected to and held to represent or be about objects without the mind and at a distance. The plausibility of the psychological account of how we see things like distance ultimately depends upon the success of the account of how we arrive at a theoretical or intelligible grasp of spatial properties. There seems, then, to be some justice in Berkeley's claim in the *Philosophical Commentaries* that the geometric theory of perception encourages a belief in mind-independent objects.

The plausibility of such a theory of distance perception based on natural geometric judgments will ultimately hang on the plausibility of the kind of causal theory of perception on which it depends. It requires we take what we see to be the causal (or in the case of Malebranche, "causal") result of the action of independent extended matter on our sense organs, giving rise to mind-dependent sensations. This sort of claim will be in the end justified when we establish that it is appropriate for a psychophysical theory such as this one to attribute the independent property of extension to the corporeal world, while accounting for sensible properties as the result of the way the mind works. This general approach, with its distinction between a corporeal world and a visual world, requires there be a principled way of distinguishing the properties that describe the corporeal world from those that are merely the result of the way we see. According to Malebranche, we can make such a distinction because we can identify what we know about extension to be derived from an intelligible idea, one from which we can draw all necessary connections, whereas what we know by means of sensation is confused, revealing no necessary connections with anything else. Malebranche's argument has as a consequence that the idea of intelligible extension, as a pure idea, is the same no matter what sensations are accompanying it. We have the same idea of extension whether we are seeing a color patch or feeling a patch of warm water. The thesis of the homogeneity of the idea of extension is therefore important to Malebranche's way of distinguishing sensible from intelligible properties, which is what permits him to identify intellectual properties with mind-independent essences. Thus Berkeley's interest in the heterogeneity of the ideas of sight and touch has relevance to this notion of Malebranche, that intelligible ideas are of mind-independent essences. In addition, the calculations Malebranche has the visual system perform, which are recapitulations of its causal history, require the supposition that the same spatial properties are shared by the mind-independent object and the mind-dependent object we see.

The result of reading Berkeley's project against the background of the theories of vision to be found in Descartes and Malebranche has the result of making possible a somewhat different account of his motivation than the one in what I earlier called the Principles account. Berkeley's argument is directed toward a demonstration that, since we do not perceive the same ideas by sight and by touch, we lack an important element in a story that links our mind-dependent sensations with some set of mind-independent properties. The focus of his attention is on theories that maintain it is possible to isolate an account of the nature of a mind-independent world. The particular project of the *New Theory* will be to contrast the defectiveness of a theory of vision that is supported by the belief in mind-independent intelligible qualities with the success of his own theory, an account of how we see distance, size, and situation by means of heterogeneous sensations of sight and touch, lacking connections with each other, either necessary or by resemblance. Berkeley's project, conceived negatively as overthrowing an essentialist account of a mind-independent corporeal world, will be successful to the extent that he has successfully carried out a positive program of explaining space perception. A crude idealism, claiming that we don't really see distance, because distance perception is "in the mind," can form no part of such a project.

PART TWO

A Commentary on the *New Theory of Vision*

4

How Distance Is Perceived by Sight

The opening sentence of *An Essay towards a New Theory of Vision* promises an account of how we actually do come to see distance. Berkeley's second sentence, however, appears to take back such a positive attitude toward the possibility of perceiving distance by sight. In it he says, "It is, I think, agreed by all that distance, of itself and immediately, cannot be seen" (*NTV* 2). It is this sentence that has been taken to indicate that Berkeley is not, in fact, prepared to give a positive account of how we see distance, because what he actually thinks is that distance cannot be seen.[1] Such a judgment, however, is too hasty. In section 11, commenting on section 2, Berkeley says: "Now from sect. 2 it is plain that distance is in its own nature imperceptible, and yet it is perceived by sight. It remains, therefore, that it be brought into view by means of some other idea that is it self perceived in the act of vision." From this section, it seems Berkeley is far from supposing the plain fact is that we can't see distance. Rather, he is saying that since we can see distance, we must look about us for something that will account for this ability.

The procedures Berkeley follows as he lays out his account of

1. For two very complete accounts of Berkeley's theory of distance perception which incorporate the assumption that Berkeley thinks we can't see distance, see Armstrong, *Berkeley's Theory of Vision*, pp. 2–22, and Pitcher, *Berkeley*, pp. 4–24. This general approach is widely shared. See also the accounts given by G. Dawes Hicks, *Berkeley* (New York, 1968), Warnock, *Berkeley*, Richard J. Brook, *Berkeley's Philosophy of Science* (The Hague, 1973), and Tipton, *Berkeley: The Philosophy of Immaterialism*, pp. 200–210.

distance perception in the introductory sections of his discussion confirm he is doing nothing more alarming than proposing that the perception of distance by sight is a multistage process. Berkeley's account falls into line not only, as he says, with that of other theorists of his day but with many accounts of distance perception current among contemporary psychologists.[2] But even on the assumption Berkeley has intended to develop a positive account of distance perception by sight, it is nevertheless clear by the end of his discussion of this problem, he thinks he has arrived at some new and surprising facts about distance. In *NTV* 45, Berkeley says: "So that in truth and strictness of speech I neither see distance it self, nor anything that I take to be at a distance. I say, neither distance nor things placed at a distance are themselves, or their ideas, truly perceived by sight." Berkeley's final account of distance perception is undeniably intended to capture some novel views about distance. Thus Berkeley is doing two things. He is putting forward an account of how we do see distance, an account that is intended to be persuasive because it can do a better job of accounting for the facts of distance perception than the theory he sets up as its rival, the geometric theory. He is also, by means of his new theory, leading us to accept some surprising facts about distance; in particular, that it is entirely nonvisual in nature. A good understanding of Berkeley's theory therefore requires distinguishing his admittedly surprising conclusions from those aspects of his account of distance perception Berkeley does not regard as surprising but instead as part of what he can take for granted, as "agreed by all."[3]

What Is Agreed by All? (NTV 2)

Berkeley's initial statement, that "distance, of itself and immediately, cannot be perceived by sight," lays down the terms of a

2. With the exception of that of J. J. Gibson and his followers, almost all contemporary work in psychology assumes that the perception of distance is indirect.

3. Pitcher, for example, cites *NTV* 45 as evidence that when Berkeley says distance is not immediately perceived by sight he means the visual perception of distance is illusory. In my view it is important to distinguish the claims of *NTV* 45, which draw conclusions based on Berkeley's own theory, from the claim about immediate perception, which serves as a premise to Berkeley's argument and which he shares with his rival theory.

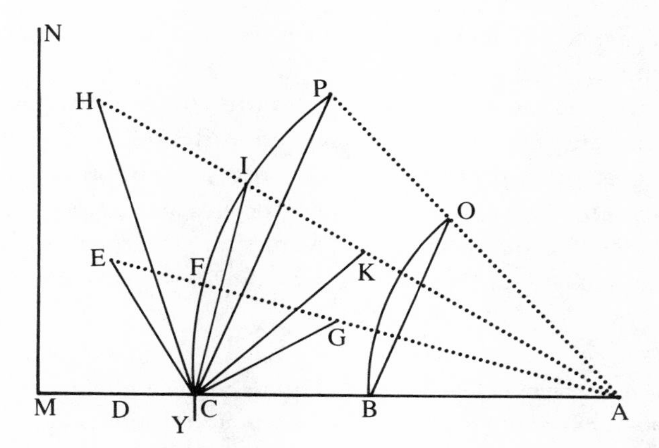

Figure 2. Redrawn from Malebranche's *Search after Truth.*

problem about distance perception, a problem he regards as widely recognized. There seems to be some justification for this opinion. Malebranche begins his own discussion of distance perception with a demonstration that, although explicitly about motion, seems to involve an attitude very similar to Berkeley's about distance perception by sight. Malebranche illustrates his claim with Figure 2, about which he says:

> Let A be the eye of the viewer, C the object, which I take to be at some distance from A. I claim: that although the object remains immobile at C, it may be thought to recede to D, or to approach B. That although the object recedes toward D, it may be thought to be at rest at C and even approach B, and that, on the other hand, although it approaches B, it may be thought to be at rest at C and even recede toward D. That although the object has advanced from C as far as E, or even H, G, or K, it may be thought to have moved only from C to F or I; and that on the other hand, though the object has moved from C to F or I, one might take it to have moved to E or H, or even G or K. (*ST*, bk. I, chap. 9, sect. 1, pp. 40–41)

Malebranche's point is that it will not be possible to tell if an object has moved along the same line of sight, because different distances along the same line of sight can't be picked up by eye. The same object can be equally well taken to be in any one of a number of different places. Berkeley could find in Malebranche, then, the view that there is a problem with distance perception by sight, because different distances can look the same.[4]

Malebranche, however, does not, in this context, give any explanation for this problem. The explanation Berkeley adopts for the existence of this problem about distance perception consists of a single sentence, which, as many have recognized, is close enough to be considered a paraphrase of a very similar account of distance perception to be found in William Molyneux's *Dioptrica Nova*. Berkeley's sentence reads:

> For distance being a line directed end-wise to the eye, it projects only one point in the fund of the eye, which point remains invariably the same, whether the distance be longer or shorter. (*NTV* 2)

The corresponding passage from Molyneux's *Dioptrica Nova* runs as follows:

> In Plain Vision the Estimate we make of the *Distance* of Objects (especially when so far removed, that the Interval between our two Eyes, bears no sensible Proportion thereto; or when look'd upon with one Eye only) is rather the Act of our *Judgment*, than of *Sense*; and acquired by *Exercise* and a Faculty of *comparing*, rather than *Natural*. For *Distance* of it self, is not to be perceived; for 'tis a Line (or a Length) presented to our Eye with its End towards us, which must therefore be only a *Point*, and that is *Invisible*. (p. 113)

In turning to Molyneux for an account of the problem of distance perception, Berkeley seems to be relying on the assumption that the problem is one that can be understood simply in terms of the optical facts. That different distances look the same, Malebranche's problem, is the result, according to Berkeley, of the way distance is

4. Hicks, although he also refers back to Malebranche, puts the problem somewhat differently. He says: "But the prior question is, by what means does it come about that the contents of the visual field are connected with objects external to the conscious subject or to his body? How does the whole visual field acquire its reference to extra-organic space?" (*Berkeley*, p. 42). I think this problem, that "outness" is not part of the deliverance of vision, is clearly important to Berkeley and shaped his discussion of the Molyneux man.

registered on the retina.[5] Because of what is on the retina, what we see is ambiguous with respect to distance itself along the line of sight. Berkeley seems to be supposing that what is agreed by all involves a set of common or general assumptions about how a particular set of optic facts supports conclusions about what we see.

The optical facts about which Berkeley could expect general agreement were those that had been basically established by Johannes Kepler.[6] The crucial optical fact is that vision is a matter of what happens when light rays transmitted from an illuminated object are focused by the lens on the back of the retina. Thus Molyneux describes the physical situation as follows:

> By the forementioned Scheme we perceive the Rays from each Point of the Object are all confused together on the pupil . . . , So that the Eye is placed in the Point of the Greatest Confusion: But by means of the Humours and Coats thereof each Cone of Rays is separated, and brought by it self to determine in its proper Point of the *Retina*, there painting distinctly the Vivid Representation of the Object, which Representation is there perceived by the *sensitive Soul* (whatever it be) the manner of whose Actions and Passions, He only knows who Created and Preserves it, whose Ways are Past finding out, and by us unsearchable. (*Dioptrica Nova*, p. 104)

The eye is a physical organ, which stands in a particular physical relationship to the object seen. The physical relationship is determined by the operation of light rays from the object focusing through the lens onto the retina. Since we see when light rays from the object are focused on the retina, an account of how the eye works is an account of how light rays affect the retina. Such an account

5. Alan Donagan has pointed out that when Berkeley gives a formulation of his distance argument in *Alciphron*, he spells out that this is a problem about the way things look. The passage in *Alciphron* reads: "Therefore the appearance of a long and of a short distance is of the same magnitude, or rather of no magnitude at all, being in all cases one single point" (*A*, *IV*, Wks. 3, p. 150). See Donagan, "Berkeley's Theory of the Immediate Objects of Vision," in *Studies in Perception*, ed. Peter K. Machamer and Robert G. Turnbull (Columbus, Ohio, 1978), pp. 312–35. In *Three Dialogues*, the problem is said to be that "a line turned endwise to the eye," which is to say, distance, cannot be perceived by sight (*3D I*, 202).

6. See, for example, David C. Lindberg, *Theories of Vision from Al-Kindi to Kepler* (Chicago, 1976).

reveals that light from a point at any distance along the line of sight will focus on the same point on the retina. That is, light from a point at any distance in the same direction will have exactly the same physical effect on the retina.[7] The distance between us and some object at which we are looking can be thought of as the length of an imaginary line, corresponding to the line of sight in the direction in which we are looking, between us and the object. Physically speaking, however, light from objects at whatever distance is projected on the retina, as Berkeley says, as a point, and light from objects at different distances but in the same direction will therefore be projected as the same point. So physically speaking, we cannot see different distances along the same line of sight.

This result, that light from points at different distances affects the retina in the same manner, will amount to the claim that distance in itself and immediately cannot be seen, if there is assumed to be a correlation between the way something is registered on the retina and the way it is seen. This correlation has formed an important element in the theory of sensation that Malebranche shares with Descartes.[8] Sensations are interesting and important because it is in sensing that we become aware of our surroundings. A sensation is a way of apprehending or representing or becoming aware of what's out there and, as such, is a useful feature of mental life. But we have sensations by virtue of having physical organs of sensation, like the organ for vision just described. The existence of the sensations

7. This is not to say it will look just the same, since light from different distances may appear brighter or fainter, etc. The geometrical effect, however, will be the same.

8. Hatfield and Epstein, in their very interesting "The Sensory Core and the Medieval Foundations of Early Modern Perceptual Theory," emphasize that what Descartes adds to earlier physiological theory is the claim that what appears on the retina has a mental correlate, the sensation, which is at one and the same time held to be in consciousness and also unnoticed. They raise as a problem why it was important to Descartes to insist on the presence of a mental state corresponding to the retinal image, given that this mental state is not what is supposed to engage our attention in perception and hence is not what we think we see. My own view is that this move is demanded by the picture of sensation as a mode of apprehension mediated by corporeal organs. The laws of the union of mind and body, in Malebranche's terms, demand that each sensation have its own particular physiological correlate. Thus Descartes and Malebranche are both prepared to go somewhat further in their discussions of the working of the sensitive soul than was Molyneux in the passage quoted above.

depends upon the proper functioning of the sense organs. Both Malebranche and Descartes hold that the peculiar nature of sensation derives from the fact that sensations are the mental representations of corporeal changes originating from external objects. The way sensation works is that alterations in our sense organs, brought about by the action of physical bodies on those sense organs, are apprehended by the perceiver as sensations, or, as Descartes says in *Le Monde*: "Our mind . . . represents to us the idea of light each time the action that signifies it reaches our eye."[9]

If sensations are just ways of apprehending physiological states, then different sensations will reflect different physiological states. Light and color, for example, are the visual sensations that constitute the way in which retinal changes are represented. Thus there is a problem with distance perception, because light from objects at different distances along the line of sight can't differentially affect the retina and hence can't affect the way in which the object is sensed. Different colors can affect the retina differentially so as to look different, but different distances cannot. It is for this sort of reason Malebranche introduced the theory of natural judgments. Our understanding of the nature of sensation tells us we apprehend through apprehending physiological changes in our sensory system. Therefore our visual sensations cannot be of distance. But we *can* see what distance an object is from us, with variable degrees of accuracy. Whenever we do make discriminations, even though these discriminations could not be made on the basis of what appears on the retina, then it is clear there has been some sort of supplementation. Our visual system has at its disposal other information. The

9. René Descartes, *Le Monde, ou traité de la lumière*, trans. and intro. Michael Sean Mahoney (New York, 1979), p. 3. The account that I am giving of Descartes's views follows that of Yolton in *Perceptual Acquaintance from Descartes to Reid*. Yolton calls the picture of sensations as the apprehension of physiological states the 'reverse-sign' relation and says of it: "It is the nature of our mind to be so made that, when movements of a certain sort occur in the nerves and brain, we have specific sensations. There are specific changes in the brain (changes of movement, of force, and of manner) that enable the mind to know. But knowing (perceiving) is not reading off from our sensations properties of the world. Perceptual knowledge is the having of these sensations. This is the point of the reverse-sign relation: ideas are not signs of things, they are the interpretations of physical motion (of things), the cognitive counterpart of things and their physical features. Interpretation is not signification; it is representation" (p. 26).

discrepancies that exist between what we see and what can appear on the retina require we take vision to be a multistage process. What we can see with respect to distance is not the same as what we can sense and so requires additional processing. Thus the physiological and optical facts of the sort Berkeley cites in *NTV* 2 are, for *Malebranche*, not a reason for saying that distance cannot be seen but a reason for saying that distance cannot be *immediately* seen, if we understand this to amount to the claim that distance itself cannot be sensed because it cannot be represented on the retina. These facts form the basis for Malebranche's view that distance perception is multistage and is only accomplished by means of various forms of supplementation, or natural judgments. This is presumably, then, the sort of position Berkeley has in mind in *NTV* 2. It is "agreed by all" that distance itself, unlike light and color, is not sensed as a result of the operation of light rays on the retina. The retina cannot respond differentially to distance and changes in distance. The problem will be to explain what else is available to the visual system to account for its ability to see distances.[10]

Immediate Perception (NTV 2)

Given this sort of general understanding of the problem of distance perception, it should be clear that the notion of immediate perception involved here is a psychological notion.[11] Immediate perception is perception that depends upon no other perception, and Berkeley usually characterizes it as perception that takes place

10. For another account of *NTV* 2, less historically oriented than mine, see Pitcher, *Berkeley*, pp. 14–20.

11. For some other discussions of immediate perception, see Armstrong, *Berkeley's Theory of Vision*, pp. 2–9, Pitcher, *Berkeley*, pp. 9–12, Georges Dicker, "The Concept of Immediate Perception in Berkeley's Immaterialism," in *Berkeley: Critical and Interpretive Essays*, ed. C. M. Turbayne (Minneapolis, 1982), pp. 48–66, George Pappas, "Berkeley, Perception, and Common Sense," also in Turbayne, pp. 3–21, and "Berkeley and Immediate Perception," in *Essays on the Philosophy of George Berkeley*, ed. Ernest Sosa (Dordrecht, 1987), George Pitcher, "Berkeley on the Perception of Objects," *Journal of the History of Philosophy*, 24 (1986) 99–105, and Kenneth Winkler, *Berkeley: An Interpretation* (Oxford, 1989), pp. 149–61.

without suggestion or inference.[12] In and of itself, this characterization has not been entirely unproblematic, since a great many of the things we perceive, including, of course, distance, do not appear to take place by means of any additional processing.[13] But the discussion of distance shows that we decide whether or not a perception is immediate not by whether it feels immediate but by examining what we know about the physiology of the senses. Immediate visual perception is determined by the workings of the visual system. It will include whatever can be represented by that system, in this case, whatever is represented on the retina. Anything we perceive by sight which cannot be represented on the retina is a case of mediate perception. Thus for each sensory system, we immediately perceive the proper objects of each sense. Berkeley, in fact, frequently describes what we immediately perceive by listing the proper objects of each sense.[14] Questions about whether something can be correctly described as being immediately perceived will be entirely a function of the details of whatever psychological theory of perception is considered to be correct.[15]

The notion of immediate perception, moreover, is not intended to provide any kind of epistemological warrant.[16] It is not the case that

12. See, for example, *3D I*, 174: "The senses perceive nothing which they do not perceive immediately: for they make no inferences." I agree with Kenneth Winkler that the claim that immediate perception takes place without an intermediary, the second sense isolated by Pitcher, amounts in Berkeley's terms to the claim that it takes place without inference.

13. See Pappas, "Berkeley and Immediate Perception," and Pitcher, "Berkeley on the Perception of Objects," on the question whether medium-sized physical objects are immediately perceived.

14. "You will farther inform me, whether we immediately perceive by sight any thing beside light, and colours, and figures: or by hearing, any thing but sounds: by the palate, any thing beside tastes: by the smell, beside odours: or by the touch, more than tangible qualities" (*3D I*, 175). See also *PHK* 1. Kenneth Winkler has ingeniously mounted an argument against Pitcher by distinguishing immediate perception from proper perception, but to my way of thinking, Berkeley's way of characterizing immediate perception limits what is immediately perceived to the proper objects of each sense.

15. Thus on some accounts of color perception, color would not be immediately perceived either.

16. For further argument that Berkeley's notion of immediate perception is non-epistemic, see Pappas, "Berkeley and Immediate Perception."

if we immediately perceive something, we are then supposed to have especially good reasons for any belief we hold on that basis, that we immediately perceive things just the way they are or even just the way they are to us. Nor should it be supposed that if we don't immediately perceive something, then our beliefs are on shakier grounds. There is no reason to suppose that because distance is not immediately perceived, then our beliefs about the distances of things, or our belief that things are at a distance, somehow lack warrant. In fact, in the way in which Malebranche uses the term 'immediate perception,' he intends us to draw the contrary conclusion. Malebranche says:

> it should be noted that there are two kinds of beings, those our soul sees immediately, and those it knows only by means of the former. For example, when I see the sun rise, I first perceive what I see immediately, and because I perceive this only because there is something outside me that produces certain motions in my eyes and brain, I judge that this first sun, which is in my soul, is external to me and that it exists. (*ST*, bk. I, chap. 14, sect. 2, p. 69)

For Malebranche, what I perceive immediately is that which I perceive by virtue of certain motions in my eyes and brain. The language of immediate perception comes into use in order to express the fact that there is a way in which perceivers represent simply by virtue of having sense organs. Malebranche wants to use this language in order to make the point that what I perceive immediately will constitute a way in which the sun is represented by me but not a way in which the sun is in itself or externally. So, for Malebranche, we believe something false when we believe that the sun is as we immediately perceive it to be, whereas our true beliefs will be about the sun we perceive, not immediately but by means of our immediate perceptions. When we talk about what we immediately perceive, we are talking about ways of perceiving or sensing that just happen, by virtue of our sensory system, and, therefore, inevitably just happen in the way that they do. But that is not at all the same as saying that we are entitled to claim special or incorrigible evidential status for what we immediately perceive.[17]

17. Berkeley does sometimes seem to be committed to the view that we cannot be mistaken about what we perceive immediately. But it is also clear that he imposes

As I have indicated, we are not expected to have any commonsense intuitions about what is immediately perceived. Since we identify what we immediately perceive by sight by discovering what the eye can do, we shouldn't expect to be able to read this off the visual field. What we immediately perceive is what we sense, that is, what we apprehend just by virtue of the kind of sense organs we have. When we know what kinds of ways of representing are available to us by virtue of having the kind of visual system we do, then we can factor this information out of the visual field as a whole. Doing so will give us a distinction between what we perceive immediately and what we perceive through the mediation of other processes. The visual field itself, the way things look to normal perceivers, is not the same as the way in which we apprehend through sensation. What sets up the kind of problem, like the distance problem, that Malebranche solves by talking about natural judgments is that what we see is other than and contains more information than what we can sense. The distance problem exists as a problem because, on the one hand, things do look to be at different distances from us, and on the other, our knowledge of optics shows us that information about these various distances can't be represented on the retina. The problem wouldn't exist unless things looked to be at different distances and hence can't be solved by reminding us that this is how things actually look.[18]

The use of the language of inference has perhaps been misleading here. Since immediate perceptions, the ones represented on the retina, just happen to us, it is of course inevitable that what we perceive immediately we perceive without making any inferences. The way the senses work means the senses make no inferences. The kind of supplementation Descartes and Malebranche provide is one that could be worked out through inference, but they both seem to think, in their discussions of the natural geometry they use to ac-

---

conditions, in particular that our perception be attentive. See Pappas, "Berkeley and Immediate Perception."

18. This is the sort of attitude that seems to underlie, for example, what George Pitcher writes about Berkeley on distance. Pitcher says: "I believe that all of Berkeley's troubles in this area could be resolved at once and painlessly, if he simply . . . admitted—what seems obviously true—that visual appearances are three-dimensional, that things just do normally look, in the full-blooded sense, to be at various distances from the visual perceiver" (*Berkeley*, p. 23).

count for some forms of distance perception, that we shouldn't expect to find such inferences taking place as a conscious, intellectual process. We shouldn't expect to find a first stage, that which we immediately perceive, followed by an inference. What we should expect to find is things looking to be at some distance from us.

If Berkeley, and Malebranche and Descartes and Molyneux and numerous others, were wrong about distance perception, this error would not be uncovered, then, simply by thinking about how things look. It can turn out that distance *is* immediately perceived only if the distance between the perceiver and the object seen is in fact something that the visual system is built to register. What this would presumably have to amount to is a demonstration that there are other effective ways that the visual system can receive information besides the focusing of light rays on the retina. Such a demonstration would be upsetting to Berkeley's general scheme of things only if he had a deep commitment to the retina as the effective means for registering *visual* information, or if he had a deep commitment to the view that we register distance in a different way than we register light and color. Neither of these is in fact the case. As Berkeley himself points out, if it turns out that we perceive distance in the same way we do color, this would only strengthen his hand. He has Philonous remark in *Three Dialogues*: "But allowing that distance was truly and immediately perceived by the mind, yet it would not thence follow it existed out of the mind. For whatever is immediately perceived is an idea: and can any *idea* exist out of the mind?" (*3D I*, 202). Berkeley's discussion of distance perception begins with the claim that distance is not immediately perceived, not because this claim is essential to his projects, one that he *had* to assent to, but simply because Berkeley, in common with the theorists he was endeavoring to refute, thought it was true.[19]

19. It is interesting that Thomas Abbott (who, writing in 1864, was, I think, one of the first to try to refute Berkeley by showing that immediate perception of distance is not impossible) also dismisses as irrelevant that there might be necessary connections between the idea of distance and variations in the circumstances being perceived. It seems that it is when the geometric theory that Berkeley set out to refute has dropped so far out of sight as to seem inconceivable that Berkeley's own project is given a new reading. See Thomas K. Abbott, *Sight and Touch: An Attempt to Disprove the Received (or Berkeleian) Theory of Vision* (London, 1864).

The Concept of Distance (NTV 2)

Misunderstandings about the nature of Berkeley's claim that distance is not immediately perceived have resulted in some confusions about the notion of distance he is using. The possibilities for confusion have been exacerbated because commentators have moved from one notion of distance to another without noticing the differences among them and hence without noticing there is actually considerable room for misunderstanding what Berkeley means by distance. D. M. Armstrong is a good case in point. When he first begins his discussion of the claim that distance is not immediately perceived, he says: "The word 'distance' gives no trouble: by 'distance' here Berkeley simply means distance out from the eye (sometimes he calls it 'outness'), that is, along the line at right angles to the retina" (p. 2). But Armstrong's proposal that by 'distance' Berkeley means 'outness' fails to distinguish between two notions other commentators have suggested ought to be kept separate. Richard Brook, for example, has pointed out that distance can mean what he calls metric distance,[20] how far away the object looks to be (the sort of thing about which I make rough-and-ready judgments by eye and surveyors make very exact ones), or it can be used to mean outness, a characteristic the visual field has when things look to be sitting out there independent of the perceiver. This second notion, which is part of the visual look of things, is a good candidate for what people have in mind when they say things just do look to be at a distance, they just do look to be out there, and so distance must really be immediately perceivable. But the problem Berkeley found Descartes and Malebranche solving geometrically is actually a problem of metric distance.[21] This problem is about how we can tell by eye the amount of distance an object looks to be from us, something we are pretty good at telling for near objects, which generally look to be at the distance away they are, and very bad at for great distances,

20. See Brook, *Berkeley's Philosophy of Science*. Phillip Cummins similarly makes a distinction between outness and what Cummins calls 'exact distance,' in "On the Status of Visuals in Berkeley's *New Theory of Vision*," in *Essays on the Philosophy of George Berkeley*, ed. Sosa.

21. Brook thinks that it is only the problem of the perception of metric distance that could even be supposed to have a geometric solution.

such as that between us and the moon, which tends to look considerably closer than it is. This problem is distinguished explicitly by Malebranche from that of outness, although, of course, Malebranche thought that neither distance nor outness was immediately perceived but was the result of different natural judgments. There are two different natural judgments involved when I take the moon to be one hundred paces away and when I take the moon to be out there, although Malebranche did think that the objects I see as out there are the ones I have seen to be at a distance, since he thought the natural judgments of outness result when I take what I see to be at the distance from me that its cause is located. Thus for Malebranche, seeing things at a distance is accompanied by seeing them to be out there, though not by any means as a result of a single or immediate perception. Berkeley too speaks frequently of objects as being "at a distance, or without the mind" (*NTV* 41; see also *NTV* 43, 50, 55, 95, and *PHK* 42). But the use of this phrase is not, in and of itself, a reason for supposing he thinks he is talking about one notion rather than two. It is certainly not a reason for supposing the concept of distance that Berkeley says is agreed by all not to be immediately perceivable by sight is anything other than metric distance.

Slightly further on, Armstrong introduces a third notion. He says: "Similarly, it seems that sometimes the bulge on the front of a tomato can be immediately seen. Depth seems to be immediately seen. Now, if we have understood Berkeley aright, he is asserting that, whatever the views of common sense, I cannot ever immediately see the bulge on the face of a tomato, that is, depth or distance is never immediately seen" (p. 5). Armstrong is equating Berkeley's claim that distance is not immediately perceived by sight with the claim that depth or bulginess is not immediately perceived by sight. Armstrong is followed by a great many other commentators in making this equation. George Pitcher, for example, also says that to say that distance is not immediately perceived entails that "visual appearances are altogether flat" (p. 8).[22] Looking to be at some distance is not exactly the same notion, however, as looking to be in depth.[23] I

22. Pitcher in fact says that Berkeley's views on distance perception ought to be understood as being about "visual depth perception" (*Berkeley*, p. 5).

23. Schwartz is the only person, so far as I know, who has pointed out this differ-

can say of the engraving on the wall in front of me that the lake in the background looks further away than the trees in the foreground, even though the engraving presents a visual appearance that is perfectly flat; and, more tellingly,[24] a picture of, say, just a globe seen through a stereoscope will look bulgy or in depth but, without any other distance cues, it won't look to be at any particular distance.[25] That both Descartes and Malebranche thought distance perception is something we are better at some times and not at others is again a reason for assuming it is metric distance they are talking about rather than depth or bulginess. There is no reason to suppose they think sometimes objects look very bulgy whereas at other times they look flat.[26] That is, most contemporary discussion of the *New Theory* has assumed that to say distance is not immediately perceived is to say that things don't look to be "out there," and if things don't look to be "out there," then, it is thought, it must be because they look flat. Thus the notion Berkeley is making a claim, and a clearly false claim, about how things look, namely that things look flat, when he says that distance is not immediately perceived, has contributed to confusions between distance, outness, and depth. The assumption that Berkeley thinks that things look flat is highly questionable, not least because Berkeley explicitly denies it, saying: "From all which we may conclude that plains are no more the immediate objects of sight than solids" (*NTV* 158). Thus there is good reason for distinguishing the concept of distance Berkeley is discussing from both outness and depth. If it is clear the problem he is addressing is how it is that, by seeing, we tell objects are the metric distance away they look to be

---

ence with respect to Berkeley, in "Seeing Distance from a Berkeleian Perspective" (unpublished).

24. Since one is inclined to say, yes, but the reason why I know that the lake part of the picture isn't really further away than the trees part is due to the lack of bulginess.

25. This is Schwartz's example. It is true that in ordinary circumstances, when something looks bulgy, viewers are able to make some estimate about how far away the front of the object is from the back, but, in the absence of distance cues, mere bulginess alone wouldn't permit such an estimate. In the example of the stereoscope, the front of the cube could be any distance at all from the back.

26. As Mill pointed out some time ago, and as Schwartz has argued since, if the notion that Berkeley is trying to explain is metric distance not depth, then facts about stereoscopic vision are not necessarily relevant to his point.

from us, then his claim that we don't do this immediately gains plausibility. He is saying that retinally registered information is insufficient to inform us whether we are seeing distances of ten or one hundred or one thousand feet.

In *NTV* 2 Berkeley says he is doing nothing more than pointing out a problem he takes to be well-understood. The problem he raises is one concerned with the visual awareness of distance. It is important to stress this is a problem about the nature of visual awareness. The problem comes up because of the way the visual system is thought to apprehend distance. It is not, at least as it is stated here, a problem about distance. There is nothing in what Berkeley has said so far to suggest he, or anyone else, thinks things are not at a distance. That is, there is nothing in what has been said so far to suggest things are only at a distance if we can immediately see these distances. In fact, the way in which Berkeley sets up the problem in terms of the physical functioning of the retina seems to presuppose there are things at distances from the eye, from which light is reflected and focused on the retina.[27] Nor is there anything in *NTV* 2 to suggest we can't be aware of the distances things are at. Berkeley has confined his attention so far just to one modality, vision. There is nothing in what he says that would preclude the possibility we can become immediately aware of distance by other means. The problem, moreover, that comes up with respect to visual awareness is strictly limited to the *immediate* apprehension of distance. We have a problem not because things *don't* look to be at distances but because they *do*. The problem is to explain how it is we are able to become aware of distance by sight, given that this information is not registered on the retina. So long as we keep an eye on the presence of the words "by sight" and "immediately," then the highly limited nature of the issue Berkeley is initially addressing becomes plain. Berkeley

27. This is sometimes raised as a problem for Berkeley. Brook, for example, says: "If the Molyneux premise is taken seriously, then Berkeley's argument presupposes the existence of that which he is later at pains to deny; that is, three dimensional space" (*Berkeley's Philosophy of Science*, p. 52). Berkeley will have problems if he later wants to deny the existence of three-dimensional space, but simply in raising this problem, he is committed only to the claim that three-dimensional space can't be apprehended immediately through vision.

never says that everyone agrees we can't perceive distance, only that we can't perceive it by sight, immediately.

The Refutation of the Geometric Theory (NTV 3–15)

Berkeley offers a solution to the problem of distance perception which he supports in part by arguing that his is superior to the received view. It is quite common to call the received view against which Berkeley is arguing "the geometric theory," but it is also true that the existence of *a* geometric theory to which his own theory can be contrasted is to some extent the creation of Berkeley himself. What Berkeley actually does is to isolate for criticism those aspects of psychological theories of vision, for example, of Descartes and Malebranche, which are geometrical in nature. As Berkeley points out, Descartes's and Malebranche's accounts of distance perception are something of a mixed bag.[28] They make use of a great many of what are called "pictorial" cues, as that we take an object to be further away when there are a great many objects between us and the one we are looking at; or if the object is very faint, we take it to be further than some very distinct object; or if it is very small, we take it to be further than some object we are seeing that is very large. A great deal of the supplementary information to which they appeal in the discussion of distance is nongeometric in nature. It is only for a small range of items quite near to the perceiver that they appeal to geometric principles to supplement the information on the retina. Thus it is not the case that either Descartes or Malebranche operates on some general principle, as that space perception ought always to be explained as a species of geometrical reasoning. Instead, Berkeley has singled out for criticism an aspect of their theories, but an aspect that he thinks has shaped their overall picture of what space perception is like.

What Berkeley does is to divide the various cues that have been

28. In *NTV* 3 Berkeley says, "I find it also acknowledged that the estimate we make of the distance of objects considerably remote is rather an act of judgment grounded on experience than of sense."

discussed as cues to distance perception into two categories. It is part of Berkeley's contribution to the theory of space perception to point out that the various ways in which the retinal information has been taken to be supplemented fall into two groups, each of which rests on a different account of how the visual system works. The kinds of cues mentioned in order to explain how we see the distance of objects that are very far off, pictorial cues, presuppose seeing distance is something learned from experience, whereas the kinds of cues that are involved in an account of how we see the distance of near objects, geometric cues, do not. Berkeley sums this up in *NTV* 5:

> Betwixt which and the foregoing manner of estimating distance there is this remarkable difference: That whereas there was no apparent, necessary connexion between small distance and a large and strong appearance, or between great distance and a little and faint appearance, there appears a very necessary connexion between an obtuse angle and near distance, and an acute angle and farther distance. It does not in the least depend upon experience, but may be evidently known by any one before he had experienced it, that the nearer the concurrence of the optic axes, the greater the angle, and the remoter their concurrence is, the lesser will be the angle comprehended by them.

Berkeley's claim is that an account of the kind of supplementary information to be attributed to the perceiver in distance perception has been justified by using one of two different kinds of reasons.

When we explain that an object is taken to be very near when the optic angle is very large, then this explanation trades on a certain kind of connection between the optic angle and the location of the object. The size of the optic angle is determined by the location of the object. It then can be argued that knowledge of the size of the optic angle is the right sort of supplementary information to attribute to the perceiver or to the visual system, because the possession of such knowledge is what is needed to solve the distance problem and identify the location of the object. But this sort of account won't do when we explain the perception of the distance of very remote objects by means of pictorial cues. There is nothing in the nature of a pictorial cue like faintness that requires that an object appear faint only if it is at a very great distance. Therefore if we typically see faint objects as being at a great distance, it can be only because, in our experience, objects seen as faint have always been a long way off. Berkeley's procedure is to discredit the first sort of explanation, which takes space perception to rest on a knowledge of geometric principles, leaving him with the second, which takes space perception to rest on experienced connections, as the only remaining tried-and-true form of explanation. Enthusiasts of the geometric form of explanation are not going to be able to corner Berkeley by asking how else space perception is to be explained. Berkeley has to show only that a kind of explanation that everyone agrees works some of the time is, in fact, the right explanation all of the time.

Berkeley's reason for rejecting the geometric theory is that perceivers are unaware of the lines and angles that form the basis of what they are supposed to know. This is knowledge which actually belongs to theoreticians of optics and not to ordinary perceivers, who can see the distance of objects quite successfully without knowing the least hint of geometry. "In vain," he says, "shall any man tell me that I perceive certain lines and angles which introduce into my mind the various ideas of distance, so long as I my self am conscious of no such thing" (*NTV* 12).[29] The nature of Berkeley's argument has not been entirely clear. He is sometimes taken to be supposing geometric theorists claim we undergo a conscious process of reasoning, which claim he is endeavoring to refute by observing that he himself undergoes no such process. But theorists such as Descartes and Malebranche don't necessarily claim such a process is one of which we are conscious. It is true that Descartes's position is somewhat equivocal, since in the Sixth Set of Replies he talks about a

29. The assessment of this argument in the literature has varied widely. Some, and Warnock seems to be an example, find it completely convincing. Tipton points out, moreover, that Berkeley's arguments proved decisive against geometrical optics. Others, such as Armstrong, find it wholly unsatisfactory. Armstrong can't figure out why we should be expected to know the basis on which we make distance judgments and says: "So, it seems, it is no real objection to the view that distance is not immediately seen, but is judged of by lines and angles to say that we are not *conscious* of doing any such thing. All that is necessary is that the lines and angles be observed (however unselfconsciously) in a rough-and-ready way" (*Berkeley's Theory of Vision*, p. 22). I cannot say that I am clear about what Armstrong has in mind.

reasoning process that is very rapid, but in the *Dioptrics*, which is more obviously about distance perception, Descartes does not claim that the natural geometry by means of which the visual system works is a conscious process. Malebranche's attitude is even clearer. He thinks we have to attribute something with the expressive power of geometrical principles to the visual system because our ability to recognize spatial properties of objects vastly outruns any conscious computational abilities we possess.[30] The only way in which we could accomplish the visual feats we do is if God endowed the visual system with a knowledge of geometry and optics. So, Malebranche's claim with respect to judgments of space perception is that "it is not we who make them, but God alone who makes them for us" (Elucidation on Optics, p. 746). If Berkeley is taken to be criticizing geometric theorists for claiming that a process takes place that he himself can't find, then his criticism misses the mark. It will come as no surprise to Malebranche or to Descartes to be told that Berkeley can't find these principles within himself, since they never supposed he could.[31]

The issue Berkeley is raising, however, is not just about whether space perceivers are conscious of things like lines and angles; rather, it is about what justifies a claim that successful space perception can be explained by the presence of lines and angles. In *NTV* 9 and 10 Berkeley gives an account, which he no doubt thought uncontroversial, of what it means to say something is perceived not immediately but by means of something else. This account rests on a claim Berkeley regards as evident: "That when the mind perceives any idea, not immediately and of itself, it must be by means of some other idea" (*NTV* 9). It is important to remark that the circumstances under which we might want to ask whether something is being perceived mediately rather than immediately fall within what in Berkeley's terms is the range of philosophy, because it is a question about the manner in which perceivers see. It is not an anatomical or optical problem but a problem that arises when we start to talk about how things look to perceivers. Situations that get dealt with by

30. Twentieth-century readers of Malebranche will be tempted to label his position "Chomskian."

31. Nancy Maull, for example, assumes that this argument of Berkeley's just misses the point.

talking about mediate perception arise only within the context of ideas. Such problems come up because perceptual ideas derive their content from sensation. Sometimes what we see has more content than what the relevant sense organs can register as sensations. And the reason this is a problem is that only what the sense organs can register is effective in determining what we see. Knowledge derived from other means, as Malebranche's example of the knowledge of the astronomer, won't make a difference to what we see. So the problem we face whenever we find we can see something we can't see immediately is to show where we came by the content of what we see, by effective means. It is not enough simply to say the *visual system* has some information, or that the visual system somehow embodies geometric principles, even if possessing that information would be enough to account for what the perceiver sees. It is also necessary to show the *perceiver* has the information in question, that the geometric information is in such a form that the perceiver can get at it. The geometric knowledge Malebranche appeals to in his explanation is to be attributed to the visual system but not to the perceiver. We are now at a particular level of explanation, however, *after* the perceiver has had a visual sensation. The physiological events of vision have given rise to a conscious correlate, an event in the life of the perceiver. Therefore it is the perceiver, and not the visual system, that requires supplementation. The relevant supplementation must be available in the form of ideas registered or represented for the perceiver by some sense organ. This has been a rather long-drawn-out way of saying what Berkeley says, that what we have perceived but not immediately "must be by means of some other idea" (*NTV* 9). The problem of mediate perception is a problem of where our ideas come from, given the nature of immediate perception. The claim that we perceive distance by supplementing the light and color we see with lines and angles has to amount to the claim that lines and angles are ways of representing available to perceivers. Unless the lines and angles are held to be among the ways in which we represent by means of our visual sense organs, that is, held to be ways in which we visually apprehend, they can form no part of how we see. And if they are supposed to be ways of representing, then it is relevant to point out that I am not aware of them as such, that they are not any part of how I see.

In fact, Berkeley's case is that lines and angles are not an appropriate explanation of how we see, not just because they don't happen to be among the various ways in which I represent but because they couldn't be. The lines and angles in question, he says, "have no real existence in nature, being only an hypothesis framed by the mathematicians, and by them introduced into optics" (*NTV* 14). Berkeley is claiming the lines and angles that enter into the geometrician's study of optics can form no part of a *psychological* theory of vision. It is, of course, quite true that lines and angles don't form part of the visual array; they are not there, like threads running through the visual field, to be seen.[32] But the geometric theories Berkeley is considering take lines and angles not as part of what we see but as an explanation for what we see. Berkeley's point is that there is no way in which lines or angles could be considered to be effectively registered as such by the visual system.

Berkeley considers two different accounts of how we see distance, one involving two eyes and one involving only one. According to Malebranche, for example, we see the distance of near objects with two eyes, by judging when the angle formed by the rays from our eyes meeting at the object is large, then the object is near to us, but when the angle is small, the object is further away. This means we need some means of judging the size of this angle at the object, and the effective change in our sensory system that allows us, or the soul, to judge this size is, Malebranche says, "the disposition of its eyes, which varies with the angle by which it sees the object, that is, with the distance of the object" (*ST*, bk. I, chap. 9, sect. 3, p. 41). Malebranche seems to be supposing we register the turning of our eyes as angles from which we compute the optic angle of the object. Berkeley's point is then that while we can draw a picture in which angles are present, tipped or turned eyes are not in and of themselves angles, and hence there is no angle present to be registered as such.

In the one-eyed case we are said to be able to judge distance according to the divergency of rays falling on the retina. Again Berkeley's point is that while we can draw the pathway along which the light travels as a line, and thus end up with a picture containing lines and angles, there is no way in which the lines and angles can be

32. This appears to be the way in which Warnock interprets Berkeley's point here.

represented on the retina. What strikes the retina strikes it as a point.[33] Berkeley is not in this particular passage taking a stand on the nature of theoretical entities in science, nor is it the case that he can be read as doubting the existence of the physical entities, such as light rays, that the geometrical opticians incorporate into their theories.[34] Berkeley's claim is only there is no way the geometric behavior of light rays can enter into a psychological theory at this point, because there is no way in which the sensory system can represent such behavior to us. Both Berkeley and the geometric theorists agree that distance is not immediately perceived, that distance is not the sort of thing that the visual system is built to register. When the geometricians try to account for distance perception in terms of lines and angles, however, they are simply compounding the problem, for lines and angles are not the sort of thing our sensory systems are built to register either. So when Berkeley says that he can't find such representations within himself, he is telling only half the story. He also thinks there is no way in which this could happen.

Berkeley's Solution (NTV 16–28)

According to Berkeley, then, the correct solution to the problem of distance perception is one in which the cues to distance can be shown to be effectively available to the visual system. Since distance itself is not immediately perceptible by sight, there must be ideas of something other than distance, but which are ideas the visual system is built to register, with which the apprehension of distance by sight can be correlated. Berkeley's proposal is that the ideas we register when we see the distance of an object are first, the muscular sensa-

33. It is worth noting that it is not clear that either Descartes or Malebranche thought we compute the divergency of the rays, although this language occurs in both their accounts. They also stress the sort of thing Berkeley mentions, whether objects appear more or less confused, as well as the eye strain we experience in trying to bring near objects into focus. That is, as mentioned before, their accounts are by no means purely geometric but contain large amounts of the kind of theorizing Berkeley favors.

34. Thus I agree with Brook against Warnock that there is no reason to take this passage as supporting an instrumentalist reading of talk of light corpuscles.

tion of turning the eyes, either in toward a near object or closer to parallel as the object gets further away; and second, the confusion, that is, the degree to which the object appears out of focus as it gets closer to the perceiver. This second cue can be replaced by a feeling of eye strain as we strive to keep in focus an object close to our eyes.[35] What Berkeley has done is to take the cues mentioned by Malebranche and then point out they can serve as distance cues even without a mathematical or geometrical interpretation. Malebranche had tried to find a way in which geometric information could be effectively registered. Berkeley's claim is that if eye turnings or confused vision are represented at all, they are represented *as such*, as eye turnings or confused vision, and not as lines and angles. The only way in which we can be aware of changes in the disposition of our eyes or changes in the state of the retina as objects get nearer is as muscular sensations or confused vision. These are the ways such changes are represented by us; they are ways of seeing or feeling that we cannot choose but to have.

Berkeley's idea is that the geometric theorists have overcomplicated the situation by thinking of the eye as a solver of mathematical problems, about which it makes sense to ask What knowledge, if the eye had it, would enable it to come to know the distance of the objects it is seeing? So Berkeley thinks Malebranche's solution derives from the wrong motivation, that of trying to get mathematical information into the visual system. But Malebranche's solution can be saved, once his cues are reinterpreted in the manner of non-mathematical cues, as occasions in which coexisting ideas suggest one another to the mind. Malebranche's sensory cues do correlate with distance, even though, unlike their mathematical representations, they bear no necessary or conceptual connections with distance. The lesson Berkeley wants to draw is that we understand the function of the distance cues when we see we have to learn to use these sensations as cues to distance. The connections between the cues and the distances of objects are learned through experience of their coexistence. In expressing this point, Berkeley makes use of the same comparison with language which Descartes had employed

35. These cues are to be supplemented with the numerous "pictorial" cues, as "the particular number, size, kind, etc. of the things seen" (*NTV* 28).

in *Le Monde* to talk about the way in which nonresembling or unconnected ideas could suggest one another, as a word's sound suggests its meaning. There is an important difference in the context in which Berkeley uses this comparison, however, since Berkeley stresses that the connection between the sound and the word is one due to "custom."[36] This emphasis on custom enables Berkeley to underline his moral: "That the judgment we make of the distance of an object . . . is entirely the result of experience" (*NTV* 20).

Berkeley is sometimes accused of double-dealing because he later admits the various cues that suggest distance to us are generally overlooked.[37] Perceivers are not aware, for example, that a perception of a very near object is accompanied by feeling their eyes turn inward. So, it is said, Berkeley has criticized the geometric account for using cues we are unaware of and then has substituted cues he himself admits we also are not aware of. But this criticism misses Berkeley's point. Berkeley is not committed to the view that we are aware of everything passing through our mind. His criticism of claims about the use of lines and angles is not just that we are not aware of them, as we must be for their use to be effective, but that we couldn't be. There is no way for the sensory system to represent lines and angles. His corresponding claim about muscular sensations and confused vision is not that these cues form part of a sequence of conscious steps leading up to a distance perception. The various cues Berkeley mentions are present in the act of seeing, and we can discover them as such if we pay attention to them. Berkeley's claim is that we can't see the distances of objects without moving our eyes or registering images that are confused or distinct. When we pay attention, therefore, we discover muscular sensations and confused vision and not lines and angles, because the former and not the latter

36. It is not at all clear that this feature of the language analogy, all-important to Berkeley, is one that Descartes intended to have carry over to the perceptual situation, since he, like Malebranche, seemed to take the sign-signifier relation as "natural." So I think Nancy Maull is wrong in her "Cartesian Optics and the Geometrization of Nature," when she criticizes Turbayne for contrasting Descartes and Berkeley, on the grounds that Descartes too used a language analogy. When the language analogy appears in Locke's *Essay*, in 2.9.9., the perceptual ideas are being connected via a "settled habit," but Locke, unlike Berkeley, does not stress that the same can be said of the connection between sound and word.

37. See, for example, Pitcher, *Berkeley*, p. 22.

are the ways of representing we cannot choose but to have by virtue of the kind of sensory systems in use.

The Barrow Illusion (NTV 29–40)

The primary function of the discussion of the Barrow illusion to which Berkeley now turns is, as he says earlier in *NTV* 15, as an additional argument against the geometric theory.[38] It is one of three cases Berkeley raises in the course of the *New Theory* in order to show that, even if we grant the geometric theory everything it needs, there will still be phenomena inexplicable on its grounds which can be readily explained on Berkeley's own account. The purpose of the discussion of the Barrow illusion is not merely to discredit the geometric theory; indeed, the case is sufficiently unusual that proponents of the geometric theory, like Barrow himself who raised it, were not severely discomforted by it.[39] The discussion also serves to underline Berkeley's last point, that it overcomplicates matters to take confused vision as an occasion for computations about lines and angles instead of straightforwardly dealing with confused vision as a distance cue in its own right.

Berkeley quotes a long passage from Isaac Barrow laying out what Barrow saw as a puzzle. If we assume that an object is seen as nearer to the perceiver when the rays falling on the eye are more divergent and further as the rays are less divergent until, as is the limit in the normal case, the rays are parallel, then we ought to suppose, Barrow thinks, that if, by introducing a curved mirror or double convex lens, we cause the rays to the eye to converge, the object should appear even further away than ever, as he says, "at a vast distance off, so great as should in some sort surpass all sensible distance" (quoted in *NTV* 29). But in fact this is not the case. In the circumstances

38. See also the very complete account of the Barrow illusion, together with its discussion by Tacquet, in Turbayne, *The Myth of Metaphor*.

39. In the passage from Barrow which Berkeley quotes, Barrow says: "But as for me, neither this nor any other difficulty shall have so great an influence on me as to make me renounce that which I know to be manifestly agreeable to reason: Especially when, as it here falls out, the difficulty is founded in the peculiar nature of a certain odd and particular case" (*NTV* 29). Berkeley thinks Barrow is being too sanguine, and that this case is "alone sufficient to bring their credit in question" (*NTV* 29).

described, the more the rays converge, the nearer the object appears. That the object would look to be getting closer to the perceiver cannot be explained using geometric principles but is what Berkeley's theory would predict. If we concentrate on what we actually see as the rays are brought to converge more and more on the eye as the object recedes, what we see is a confused or blurry image, just as we do when we see an object getting very near. This is because, as Berkeley later explains, the rays are spreading out to the retina from a focal point in front of the retina in the case of converging rays, whereas in the ordinary example of near vision the rays are spreading out from a focal point behind the retina. Therefore, no matter which pathway the rays have taken, what we see is the same. So, Berkeley argues, we take the confused image as a sign of nearness, just as we ordinarily do, and thus the object is seen as getting closer. The upshot of the Barrow case is that the rays and the angles they form are demonstrably irrelevant to vision. Whatever is effective will be discovered in visual experience itself, in what we see.

The problem Berkeley is addressing in sections 2–40 of the *New Theory* is concerned only with the nature of the visual apprehension of distance. That there is a problem in the way in which we see distance does not in and of itself imply that things are not at distances. Nor does Berkeley's solution to the problem appear to carry with it any such implication.[40] His argument is simply that the ability to tell by sight the different distances things are from the perceiver is the result of sensations immediately available in the act of seeing which correlate with different distances. We learn to see which distances things are at when we form the habit of correlating the distance cues we immediately apprehend with the idea of the distance the object is from us. Berkeley's project so far has been to discredit a theory that took distance perception to be a kind of calculation and to replace it with one that sees distance perception as an ability, acquired through experience, to correlate with distance, cues that are conceptually unconnected with distance. The language

40. Tipton also makes the point that Berkeley's solution to the distance problem proceeds on the assumption that the things seen (visible objects, to give Tipton's language) are at a distance. See Tipton, *Berkeley: The Philosophy of Immaterialism*, p. 206.

in which Berkeley lays out this theory implies not that the things we see are not at distances but that they are. He says, for example, in *NTV* 17:

> because the mind has by constant experience found the different sensations corresponding to the different dispositions of the eyes to be attended each with a different degree of distance in the object, there has grown an habitual or customary connexion between these two sorts of ideas.

The terms in which Berkeley has expressed the distance problem assume the existence of objects disposed at different distances from the perceiver. The solution he proposes says that perceivers learn to correlate cues related to vision with their idea of the distance of the object. Solving the problem of how we *see* distance assumes we already have an idea of the distance of objects, presumably acquired by other means. There is nothing so far in Berkeley's discussion of the distance problem that could trouble anyone who is convinced the things we see are at distances from us. There seems no reason to fear that adopting Berkeley's approach to the problem would lead to a view that things are not at a distance because they are in the mind. He has so far been interested only in showing what sorts of experiences, admittedly, of course, in the mind, can account for the visual perception of distance.

5

Distance and the Issue of Heterogeneity

In the remainder of the discussion of distance perception, sections 41–51 of the *New Theory of Vision*, Berkeley explores some of the consequences of his solution to the problem of how we see distance. If there is nothing challenging to common sense in Berkeley's treatment of the distance problem, the same does not appear to be the case with respect to its consequences. The consequences Berkeley proceeds to draw from his solution initially appear quite shocking. In quick succession, Berkeley makes a number of claims, all of which have been taken to be at best an affront to common sense if not downright incredible. He begins by remarking that if a man born blind could be made to see, he would have no idea of distance by sight but would take everything that he sees to be "in his eye, or rather, in his mind" (*NTV* 41). It soon develops that Berkeley thinks that, in fact, it is the beliefs of such a person that are correct. The sighted would be mistaken if they thought that what they saw was at a distance (*NTV* 43). We will grant these beliefs are mistaken because we recognize that when I say I see the moon, neither the moon I see, nor anything like it, can be at a distance (*NTV* 44). In fact, "neither distance nor things placed at a distance are truly perceived by sight" (*NTV* 45). The final consequence Berkeley draws is that "a man no more sees and feels the same thing than he hears and feels the same thing" (*NTV* 47), because "we never see and feel one and the same object. That which is seen is one thing and that which is felt is another" (*NTV* 49).

This consequence in particular, that we never see and feel the same object, is frequently singled out as an example of Berkeley's willingness to fly in the face of common sense. It is important, however, to keep in mind at this point that Berkeley remains interested in showing how the consequences of his own theory differ from what he has presented as its rival, the geometric theory. The consequence of the geometric theory that to Berkeley was most troubling was the view that the deliverances of the visual system are at best an imperfect representation of the actual spatial properties of the external world. The consequence of his own theory to which Berkeley is beginning to call attention contributes to a different sort of picture, one that sees the various sorts of sensory information we have available to us as fully informative with respect to the natural world. The sorts of consequences Berkeley is drawing are not intended therefore as an attack on a common-sense belief that what we perceive is the real world. He is hoping eventually to show that, properly understood, our senses provide an adequate basis from which to acquire knowledge of the spatial properties of the world.[1]

The Molyneux Man (NTV 41–42)

The difficulties in understanding the nature of Berkeley's project begin with the discussion of the man born blind,[2] whom I will call, following Phillip Cummins, the Molyneux man.[3] Berkeley asks us to consider the position of such a person when suddenly made to see. We are supposed to think about what the experience of someone would be like, who, accustomed to getting about by using the sense

1. For another account of *NTV* 41–51, see Cummins, "On the Status of Visuals in Berkeley's *New Theory of Vision*," pp. 165–94. Many of the passages I discuss in this chapter are also discussed by Pitcher, *Berkeley*, pp. 25–34, who by and large takes them to be the incredible results of Berkeley's reliance on a "place-picture" of the mind.

2. There is a vast literature on the Molyneux problem. See, for example, John W. Davis, "The Molyneux Problem," *Journal of the History of Ideas*, 21 (1960) 392–408, G. N. A. Vesey, "Berkeley and the Man Born Blind," *Proceedings of the Aristotelian Society*, 61 (1960–61) 189–206, Desirée Park, "Locke and Berkeley on the Molyneux Problem," *Journal of the History of Ideas*, 30 (1969) 253–260.

3. Cummins, "On the Status of Visuals."

of touch, is suddenly able to use his sight. How would he at first interpret or make use of what he now can see? Berkeley's introduction of the Molyneux man into the discussion looks as though he is proposing a crucial experiment to test his theory: if we could find such a person and he couldn't, on first being made to see, tell how far away things are, then this will show Berkeley is right. But in fact it does not seem as though Berkeley regards his theory as requiring empirical confirmation in this manner. Instead, he thinks of the Molyneux man as the subject of a thought experiment. Given that Berkeley has already shown, against the geometric theory, that the visual perception of distance is a matter of correlating conceptually unconnected cues with distance, there is no way a Molyneux man would be in a position to perceive distance by sight, when first made to see.

The situation of such a Molyneux man had been introduced into discussions of vision by William Molyneux in a letter to John Locke and subsequently published by Locke in the *Essay* (2.9.8).[4] About the Molyneux man both Locke and Molyneux had intuitions that Berkeley's argument supports. They both thought the Molyneux man would not be able to tell which object was a cube and which was a globe, upon first being shown these objects, even though he could recognize each by touch. Berkeley is using their intuitions about the Molyneux man to undermine a view Locke shared with Descartes, that we perceive the same spatial properties by sight and touch. Descartes, in the *Dioptrics*, had introduced a comparison between the blind and the sighted in order to refute the view that what we see is an image or simulacrum of the physical world spread out before us. He argued that we can see distance through the calculation of angles, even as a blind man can work out how far away something is

4. Park, in "Locke and Berkeley on the Molyneux Problem," has pointed out that the original version of the Molyneux problem appeared in a letter of July 7, 1688, where Molyneux wrote as follows: "A Man, being born blind, and having a Globe and a Cube, nigh of the same bigness, Committed into his Hands, and being taught or Told, which is Called the Globe, and which the Cube, so as easily to distinguish them by his Touch or Feeling; Then both being taken from Him, and Laid on a Table, Let us Suppose his Sight Restored to Him; Whether he could, by his sight, and before he touch them, know which is the Globe and which the Cube? Or Whether he could know by his sight, before he stretched out his Hand, whether he could not Reach them, tho they were Removed 20 or 1000 feet from Him?"

that he is prodding with two sticks by thinking about the angles of his hands. Descartes's point is that in each case what we sense need not resemble the physical world about which we are gaining information, but his argument presupposes we can learn about distance through either sight or touch in exactly the same sort of way, using the same calculations. Berkeley thinks if you take the position of Descartes (and, following Descartes, Malebranche) that the apprehension of distance for both the blind and the sighted is a matter of the same natural calculations about the same concepts, then you will want to say the Molyneux man will be able to tell where the objects he is seeing are when he first regains his sight. Thus anyone whose intuitions are with Molyneux's and Locke's, that the Molyneux man will have trouble seeing the distances of things, must also think with Berkeley that the means of telling distance by sight are different from the means of telling distance by touch. Berkeley can use the intuitions of Locke and Molyneux about the Molyneux man to undercut their conviction that there are common ideas of things like distance perceived by both sight and touch.[5]

To make this point about the difference between the way we perceive distance by sight and by touch, however, Berkeley need only claim "that a man born blind being made to see, would at first have no idea of distance by sight." In fact, however, he goes on to claim something else: "The sun and stars, the remotest objects as well as the nearer, would all seem to be in his eye, or rather in his mind" (*NTV* 41). In discussing the man born blind, Berkeley routinely moves in what has seemed to be an entirely unjustified manner from claiming that what the Molyneux man initially saw would not seem to be at a distance to the claim that what he saw would seem to be "in the mind." This move seems to require two puzzling steps. Berkeley frequently speaks of the objects of sight as being neither "without the mind" nor "at any distance" (*NTV* 50; see also *PHK* 43). The relationship between these two phrases is not entirely clear. It is often assumed, however, as mentioned above, that Berkeley took them to be synonymous. That is, it is assumed that Berkeley supposed that if any object does not appear to have a determinate

5. For further discussion of this matter, see my account of Berkeley's treatment of common sensibles, in chapter 10 below.

spatial location, does not appear at a distance, then it does not appear to be at any place outside the mind, does not appear without the mind. This assumption unavoidably suggests Berkeley took the mind to be a kind of place things can or can't be outside of, so that he might suppose it is impossible for an object to be both at a distance and in the mind.[6] Thus when Berkeley makes the further claim that objects of sight are "in the mind," he is assumed to be imagining whatever does not have an external spatial location, be "without the mind," must have an internal, mental location, be "within the mind." Finally, of course, since Berkeley also uses the phrase "in the mind" to mean "perceived by the mind," he ends up thought to be guilty as well of equivocation. He is supposed to have tricked himself into arguing that since objects at a distance are outside the mind, they can't be perceived by the mind.[7] The plight of the unfortunate Molyneux man remains, on this account, somewhat unclear as well. He is supposed to be experiencing things as "in his mind," but what sort of an experience is it to see things as being "in the mind" and how does it differ from the experience of the rest of us, who have presumably learned to see things differently?

Within the terms of Berkeley's theory, there is a perfectly good sense, which requires no complicated metaphysical presuppositions, in which the things the Molyneux man sees would not seem to be at a distance from him. Before the Molyneux man had been made to see, he, like everyone else, was surrounded by objects located at different distances, but, unlike the sighted, he was able to tell the distance an object was from him only when he did something like reaching out and touching it. If he is now made to see, this means that when light from the objects at various distances around him is focused on his retina, he has sensations of light and color. Physiologically speaking, what he now immediately sees is the same as what everyone else sees. But he hasn't yet learned to correlate the various visual cues he is now receiving with distance, so he has no way of knowing that, for

6. Pitcher spells out very clearly the kind of reasoning Berkeley would have to be assuming here. A similar account can be found in Armstrong, *Berkeley's Theory of Vision*.

7. See Armstrong, *Berkeley's Theory of Vision*. A. R. White makes a similar charge in his "The Ambiguity of Berkeley's 'Without the Mind,'" *Hermathena*, 83 (1954) 55–65, which contains a very complete discussion of Berkeley's use of this phrase.

example, the very blurry patch taking up a lot of space in his visual field is very close to him. He has no idea at all where it is, because he does not yet know how to understand distance cues. There are no particular distances at which objects in his visual field look to be. There are distances, of course, at which the things he is seeing are located, which he could discover by touch, but the Molyneux man can't work out where these distances are by sight. The question Where is that object you are seeing? makes no sense to him, just as it would make no sense to show someone a stereoscopic picture of a cube, which otherwise has no distance cues in it, and ask How far away is that cube pictured as being?[8]

What the Molyneux man sees doesn't seem to him to have spatial location, and so, according to Berkeley, what he sees seems to him to be in his mind. The faulty or peculiar reasoning often ascribed to Berkeley's Molyneux man has him trying to find a location for what he is seeing. When he can't locate the things where they ought to be, "out there," he is then assumed to cast about for another possible location and thus to hit on the notion that things that are not "out there" must instead be "in here." But this sort of procedure assumes that we need a special sort of reason or argument for taking the things we perceive to be in the mind. Being in the mind is a kind of second-best location where the things we perceive will end up if something goes wrong and they can't be located outside the mind. There are, however, ways of taking the phrase "in the mind" in which what we would expect is that the things we perceive are in the mind, whereas to show they exist outside of the mind requires special argument. What we see is always in the mind because seeing is a way of apprehending by a perceiver, so what we see is always perceived by ("in") the mind. In this sense, there is no interesting question about why we take what we see to be in the mind, since there is no supposition that we can see without using our minds. The only interesting question is Why do we take the things we see to be perceiver-independent? that is, to have an existence without the mind, in the Latin sense of *sine*, not in the Latin sense of *extra*.[9]

This is the question Malebranche is addressing when he asks

8. This example is discussed by Schwartz, "Seeing Distance from a Berkeleian Perspective."

9. White, "The Ambiguity of Berkeley's 'Without the Mind,'" distinguishes the two available uses of 'without the mind' in this way.

about the origins of our natural judgment of "outness." Malebranche sees this as a question that needs answering because the external existence of what we see cannot be a part of our immediate experience, the deliverance of the senses. As visual sensations, they are unquestionably ways of perceiving, that is, perceiver-dependent or "in the mind." It is significant that, on Malebranche's theory, Berkeley's Molyneux man, who is unable to tell the distances of things, would lack an important part of the mechanism by means of which we make the natural judgment that the things we see have "outness," or mind-independent externality. For Malebranche, part of what is involved in taking the things I am seeing to be external is that I identify the distance at which I perceive them with the mind-independent location in space in which the causes of what I see are to be found. If we suppose Berkeley, too, took externality to be a matter that can be separated from distance, then when he talks about how the Molyneux man would be unable to judge the things he sees to be "at any distance, or without the mind," he would have two different judgments in mind. Because the Molyneux man can't take the things he is seeing to be at distances, he doesn't take them to be mind-independent. He lacks the grounds that would lead him to believe what he is seeing would continue to exist at a point in space whether it was being perceived or not. Without these data, he would take the things he sees to be what, according to Malebranche, they actually are, namely, perceiver-dependent sensations. Berkeley's use of the phrase "in his eye, or rather, in his mind" suggests he has a similar sort of conception in mind. The Molyneux man takes what he is seeing to be ways of feeling or apprehending, dependent upon the nature of his own sensory apparatus. That is, according to Malebranche, what we see is the product of natural judgments that convert mind-dependent apprehensions of color into the perception of something at a distance and mind-independent. According to Berkeley, the perception of distance is acquired. Therefore what we are born with, or what the Molyneux man has on first being made to see, is just that way of apprehending which is the result of the visual system.[10]

10. My way of understanding *NTV* 41 differs from that of Cummins in "On the Status of the Visuals" because I take the argument of this section to rest on the claim derived from Berkeley's solution to the distance problem, that distance perception

The Nature of What Is Seen (NTV 43–45)

The function of the discussion of the Molyneux man is to provide a thought experiment that allows the reader to consider the nature of what we see before we have learned to see things like distance. Berkeley proceeds to develop an account of what learning to see distance is like. The argument he is using can again best be understood by placing it against a kind of account, like that of Malebranche, which is supported by a geometric theory of vision. It is Malebranche's position that the natural judgments by virtue of which things are seen to be at a distance and external to us are all false. This claim is part of his argument that a knowledge of the natural world requires intelligible ideas and that the information provided by the senses is systematically misleading. Berkeley's account of how we learn to see the distances of things does not have this consequence, that what we learn when we learn to see the distances of things is false. Berkeley can therefore claim to have shown we have sensory-based means of coming to know spatial properties. Berkeley has proceeded to make his case by refuting a Malebranche-like position and thus by showing that learning to see distance is not a matter of coming falsely to believe what we see is at a distance. This refutation requires him to show that the things we believe to be at a distance are not the same as what we see. What we see is the nonresembling sign for what is at a distance, not a false representation of what is at a distance.

To Malebranche, what we see is a false representation of what actually exists externally at a distance because what we see is an idea of the intelligible extension that is at a distance, which has been particularized and made sensible through color. The way in which what we see is represented contains features, like the color and the particular size of what we see, which are due to the kind and the perspective of the sensory system that is doing the representing. A true representation of what is external could involve only intelligible attributes, would describe what is external only as ways of being

---

must be learned. Thus I do not take the absence of distance information in visual sensations to follow directly from the mind-dependent nature of sensations, and so absolve Berkeley of equivocating on the phrase "in the mind."

extended. Thus Malebranche's picture of what is happening when I see the sun is that I am having an immediate apprehension, which is perceiver-dependent and which I can describe as seeing something small and yellow. This immediate apprehension is of an object external to me that has the perceiver-independent property of extension and bears certain kinds of geometric relations to what I am seeing.[11] To the extent that what I see when I immediately apprehend the sun is other than what I take to be at a distance, what I see when I see the sun as being at a distance is false.

Berkeley's first move after his discussion of the Molyneux man is clearly directed at such a theory. He begins by saying that just as the Molyneux man would take what he is first seeing to be perceiver-dependent sensations, so perhaps some sighted persons can be found who are not "irrevocably prejudiced on the other side, to wit, in thinking what they see to be at a distance from them" (*NTV* 43). There are at least some sighted, namely, theorists of vision, who think at least some of what we see, in particular, color, is mind-dependent. Berkeley's argument, which will recur at much greater length later in his writings, has the function here of introducing the notion of visible extension. He wants to deny it makes sense to think of what we are seeing as an amalgam of something mind-dependent, something that exists as a way of apprehending, like color, and something mind-independent, like extension, in the way that Malebranche does when he calls the apprehension of a particular extent of color an intelligible idea rendered sensible through color.[12] Berkeley's claim here is that color can't be apprehended except as extended, so that visible extension must be understood as part of what is involved in seeing color. His case is that there is no way in which we could apprehend color, which is agreed to be mind-dependent, except as taking up a certain extent of the visual field, except as having visual extension. We can't think about or describe the apprehension of color without thinking about or describing a color

11. See Malebranche's account in *ST*, bk. I, chap. 14, discussed above.

12. The thrust of *NTV* 43 is to deny that visual extension can be conceived as mind-independent. The claim is that visual extension, like color, is mind-dependent, not that visual extension is like color in not being at a distance. If visual extension is not mind-independent, however, then it cannot be located at some mind-independent spatial distance.

patch. Thus visual extension, the extension of the color patch, is a part of the mind-dependent immediate apprehension of sight, part of the way in which we see color.

Introducing the notion of visible extension, moreover, highlights the fact that there are ways of seeing things, that take up a certain amount of space in the visual field and can be described as large or small. Berkeley's claim is we do not, even "in common discourse," confuse the extent taken up in the visual field with the extent of what is at a distance. I do not suppose, because a color patch takes up very little of my visual field, that what is at a distance is very small. This is Berkeley's point when he says, "the immediate objects of sight are not so much as the ideas or resemblances of things placed at a distance" (*NTV* 44). He is not claiming we apprehend special little objects of sight which are other than the objects sitting in space at a distance. His argument, it should be noted, does not involve the claim that the immediate objects of sight are in the mind and the objects at a distance are not. His point is simply that we do not take the way in which we apprehend the things at a distance to be an exact replica or copy of what is at a distance. We do not suppose the object will look the same when we get close. The discussion of the apprehension of distance by sight requires us to focus on distance cues, the ways things look and feel when we judge them to be at a distance. Berkeley is calling attention to the fact that what we take to be cues to distance, as that the object looks very small, are different from the properties we ascribe to the object at a distance.

The sort of claim Berkeley is making in *NTV* 44 about moons and towers is supposed to be something anyone would be prepared to grant and is not itself supposed to embody any surprising ontological ideas about what we see. Anyone would, I think, be prepared to say moons and towers look different from different distances, and anyone would find it most unusual to advance upon a small faint object and to discover upon reaching it that it is still small and faint. No one expects these sorts of observations to have as a consequence that it is not moons or towers that we are seeing, although everyone will be prepared to grant that a moon or a tower can be seen only from some aspect. These claims about the way moons and towers look at a distance would also be familiar claims to Descartes and Malebranche, and indeed, claims like these form an important part of their arsenal. They too want to stress that the content of what we see isn't just a function of how the object is but also depends upon facts about the perceiver. From this *they* want to claim the way in which we see objects must be understood to be other than the way the object is. Malebranche holds the external object that is at a distance is in its own nature imperceivable, because in its own nature it is to be characterized by mind-independent properties. According to Malebranche, what we see embodies information about what is out there but in an imperfect form. What we see is intelligible extension rendered sensible. The notion that what is out there can be characterized by mind-independent properties that can be abstracted from what we see has as a consequence that our eyes always deceive us. Our eyes show us extension under an aspect that is always other than the extension of the external object. Thus it is possible to find incontrovertible claims about the variability of sense perception leading to a conclusion that we never see what is at a distance. Such conclusions, however, must be supported by views about the nature of the external object which it is Berkeley's project to refute.

Berkeley follows his remarks in *NTV* 44 about what seeing objects at a distance is like with an account of how visual perceptions are supplemented by tangible perceptions. He is not arguing that we are mistaken when we suppose, on the basis of what we see, there are objects at a distance. We would be entitled to conclude, from the fact that visual apprehensions are not immediately of distance, that objects are not really at a distance only if we had no other way of apprehending distance than by sight. Berkeley's point is that this is not the case. We have a faculty by means of which we can immediately apprehend distance and things placed at a distance, namely, touch. The culmination of Berkeley's argument about distance is to be found not in the often-stressed passages in *NTV* 44 about the variability of visible perceptions but in the account in *NTV* 45 of the uses to which we put tangible ideas. Berkeley's position is that the object at a distance and its distance from us are perfectly perceivable. The object and its distance are not what we see, because we apprehend them by touch and not by sight. Berkeley's account of how we learn to see the distances of objects is that we learn to correlate the various visual distance cues with the distances of objects. We come to associate ideas acquired in vision, which in their own nature have nothing to do with distance, with the idea of distance. Berkeley believes his account requires we learn about distance somehow. His claim is we are able to apprehend distance kinesthetically. I have an idea of the distance something is from me when I perceive how long and at what speed it takes me to get to it, or how long with what effort I reach for it and what it will feel like when I have touched it. We have at our disposal the means to perceive distance and objects at a distance immediately. These means consist of various (what Berkeley calls) tangible ideas. If I see a very small object as very far away, then I am taking its visual appearance, its looking very small, which in and of itself has nothing to do with distance, to be a sign that it will take a great deal of time to get to that object. That it will take a great deal of time for me at one location to get to an object in another location is what being very distant means. If indeed it does take me a long time to get to the object I have seen as very small, then my judgment that the object is far away is true. My seeing the object as very small is not a false representation of the size of the object but is a perfectly good reason for taking the object to be at a distance.

Since the kind of tangible ideas that do represent distance are not the same as what I see, Berkeley says that "in truth and strictness of speech I neither see distance it self nor anything that I take to be at a distance" (*NTV* 45). But this is in service of an argument that is intended not to undermine but to bolster our faith in our ability to tell that objects are at a distance. Berkeley holds that when I reach out and touch the mug sitting on my desk, I have immediately apprehended its distance from me. There is a way things feel that constitutes the distance of an object. Tangible or kinesthetic experiences are different from visual experiences. They involve different sensory systems. What we experience when we look at an object, using our eyes, is not the same as what we experience when we feel it, using our hands. We learn differently about an object by seeing it and by touching it. If what we see and what we feel are two different kinds of sensory information, then, since distance is a way things feel, we shouldn't expect to be able to see distance. Seeing something small and faint is not a misleading version of what a tower feels like. The way things look are effective cues to another accessible kind of information, the way things feel.

The Objects of Sight and Touch (NTV 46–50)

Berkeley hammers the point home with an example that is apparently going to involve revision in our "common speech." This example, however, has usually been interpreted as making a claim that quite dramatically contradicts anything we ordinarily think we are saying. Berkeley talks about the different experiences we have, using different sense modalities, of a physical object, a coach, and concludes, "we never see and feel one and the same object" (*NTV* 49). Berkeley is often taken to be arguing we should cease to suppose we are seeing and feeling the same coach. But this is not the claim he is making. The one and the same object we never both see and feel is not a physical object, a coach, but an object of sense or way of perceiving. Berkeley introduces the example of the coach with the particular goal of comparing the way we see distance with the way we hear it. He reminds us that we speak of seeing, hearing, and touching the same object, the coach, even though we don't suppose that this way of speech means the data provided by one modality are the same as the data provided by another. What we hear, what we immediately perceive by virtue of our auditory system, is variations in sound. Variations in sound, from soft to loud, say, are not what we mean by distance, nor do they have any conceptual connection with distance. So in itself and immediately, distance, as Berkeley thinks anyone would agree, cannot be perceived by hearing. Berkeley is introducing the example of the coach to remind us that even though we say, I hear the coach getting closer, we are not misled into thinking that we hear anything other than sound, even though we hear it as the sound of something approaching. We don't, moreover, conclude the sound we hear is anything like the distance we take ourselves to be learning about through hearing. Berkeley's point is not that we don't both hear and touch the same coach. To the extent that ideas from different sense modalities are observed to go constantly together, it can be assumed that in fact Berkeley thinks it is perfectly appropriate to treat them as one thing. His point is that we don't both hear and touch the same idea, that of distance, because we don't really hear distance at all. We hear loudness and feel distance.

It is this conclusion, which Berkeley regards as incontrovertible with respect to hearing, that he wants to draw with respect to seeing as well. Nobody supposes they both hear and feel the same thing, not because they think they can't hear the coach they feel but because they hear sounds and you can't feel sounds. Similarly, you can't feel colors, even color patches. Each sense modality has its own characteristic set of objects. This is only to say, however, that the way of perceiving that is visual in nature is other than the way of perceiving that is auditory or tactile. When Berkeley speculates that we have been misled in our use of language into thinking we both see and touch the same thing, he does not have in mind our tendency to say we can both see and touch the same coach. He has in mind instead, as he says, our habit of describing the immediate objects of both sight and touch as extended, figured, and moving, which he thinks leads us astray with respect to sight and touch as we are not with our other sense modalities. The way in which we see something is as different from the way in which we feel it as it is from the way in which we hear it, even though we call what we see and what we feel, but never what we hear, by the same name. That I describe a visual experience as that of seeing a cube and describe a tangible experience as feeling a cube does not mean there is some common content or way of perceiving I apprehend by both means. This is not to deny that these different and many other sensory ideas may not have been observed to go constantly together, so that there is a sense in which I am indeed experiencing the same thing, the cube, even though what I am experiencing with each sense modality is different.

That we use the same words to describe both a way of seeing and a way of touching can be misleading because it causes us to overlook the fact that when I talk about what I see (as, I see a circle) there are two different uses for this phrase. I could just be reporting a visual experience. After all, whenever I talk about seeing, there must be something going on that is purely visual in nature. So I could just be reporting on a particular way of seeing, an arrangement of light and color that is apprehended as circular. This sort of report is a report on what I can immediately perceive by sight. More often, however, I am not making such a report when I talk about what I see, because Berkeley thinks we are not all that interested in what we can learn simply through seeing. So more often, when I talk about what I see, I am actually reporting on what the object I am seeing will feel like. I am talking about the tangible ideas suggested to me by what I see. So

'I see a circle' can mean 'I see something that will feel the way a circle characteristically feels.' Confusion arises if I suppose the tangible information about the circle I touch is somehow incorporated into or is a part of the way in which I see instead of suggested by what I see, because then I lose sight of the fact that often when I talk about what I see, I am not talking about visual experiences at all. This is precisely what has happened in the case of distance. Although we say we see distance, what this means is that distance information is tangible information suggested by what we see.

Part of the reason why it is difficult to understand the nature of Berkeley's claims is that it is difficult to see exactly what the experience he is describing here is like. What is distinctive about Berkeley's theory is the role played by tangible ideas. Berkeley's solution to the problem how it is possible to perceive distance by sight doesn't incorporate any supplementary processes into the visual system at all, over and above what we immediately perceive. Seeing is seeing. The primary and immediate objects of sight are characterizable only by visual properties. They are mind-dependent apprehensions of patches of light and color, which as such can't appear to be at any particular distance from the perceiver.[13] Since this is all the visual system does, the apprehension of distance is not a deliverance of the visual system at all. It is not the visual system that works out where the object we are seeing is. We learn where the object is only through the use of kinesthetic or tactile information. Thus when we say we see distance, we are supplementing the visual system by information from another sense modality. Ideas apprehended as a part of the act of seeing serve as distance cues suggesting the distance the object is from us. The primary objects of sight are said to suggest secondary objects. The distance suggested bears no conceptual connection with what is seen, being entirely the deliverance of touch. But what does it mean to say the distance at which the object is seen is really a matter of suggestion and that what is suggested is tangible in nature? If Berkeley means, for example, that when I look at an object at a distance, I recognize its distance because I feel myself gliding

13. This is the way in which I have argued it is appropriate to interpret Berkeley's claim, repeated in *NTV* 50, that the immediate objects of sight "neither are, nor appear to be, without the mind, or at any distance off."

forward, then he is surely saying something false. The experience of seeing the distance of objects remains nothing but visual. So when Berkeley says seeing distance is a matter of suggestion, that tangible information is suggested, he can't mean I am actually undergoing a kinesthetic or tactile experience.[14]

The Language Analogy (NTV 51)

It is possible to make sense of Berkeley's proposal by taking his use of the language analogy very seriously. The tangible ideas of distance suggested by the visual distance cues are the meaning of those cues. The Molyneux man, for example, who has not yet learned which tangible distance ideas are associated with what he sees, does not yet have a visual experience that is meaningful. Although his visual experiences are the same as those of the ordinarily sighted, the Molyneux man hasn't yet learned what they mean. His experience is like that of listening to a conversation in a language you don't understand. The difference between listening to a language you don't understand and listening to one you do is that in the latter you understand what the words mean, whereas in the former you just hear noise. But, of course, understanding what the words mean is not a matter of hearing noise and then undergoing, even in imagination, a set of experiences that constitute the meaning. You hear the words as meaningful. Nor is hearing the word as meaningful the same as considering the meaning of the word. In fact, it is really only when, perhaps, someone asks you what the words mean that you actually go to the trouble of explicitly thinking about a meaning that could be said to be appended to the words. Even in this case, you certainly do not relive the meaning you are thinking about.[15] Berke-

14. This kind of objection was put forward, originally perhaps, by Abbott in *Sight and Touch*.

15. Turbayne, in *The Myth of Metaphor*, puts the point like this: "Although we may learn to understand many words by building up associations, we must not think that every time we understand a sentence we must form a mental proposition answering to the verbal, whether this proposition is taken to be either a content of intellect or imagination, a conglomerate abstract idea, or a series of images. We understand the sentence without 'cashing' its word in things or in images although we can do this and often do" (p. 99). One undoubted difficulty readers face with Berkeley's language

ley presumably intends his notion of suggestion to work in this way. He describes the analogy as follows:

> No sooner do we hear the words of a familiar language pronounced in our ears, but the ideas corresponding thereto present themselves to our minds: in the very same instant the sound and the meaning enter the understanding: So closely are they united that it is not in our power to keep out the one, except we exclude the other also. We even act in all respects as if we heard the very thoughts themselves. So likewise the secondary objects, or those which are only suggested by sight, do often more strongly affect us, and are more regarded than the proper objects of that sense; along with which they enter into the mind, and with which they have a far more strict connexion, than ideas have with words. (*NTV* 51)

Just as we hear words as meaningful, so we see the object as at a distance. We have learned what objects reached through different kinesthetic experiences look like. There is not an experience of seeing, followed by a tangible meaning. The tangible ideas are not appended to the visual experience of distance any more than the meaning is appended to the experience of words as meaningful. On the other hand, what make the experience of distance meaningful are tangible ideas. If I had never had the appropriate tangible or kinesthetic experiences, I would be unable to understand my visual experiences in terms of distance.

Berkeley's Project

Berkeley's account of distance perception has a primarily positive thrust. He supposes that, as I look about me, things do look to be at different distances, some closer, others further off. This is because the way things look has been connected in my mind with the experi-

---

analogy is that the phenomena he refers to, though very familiar, are not taken to be particularly well-understood. Hearing a word meaningfully is an experience everyone has had, and thus it is possible to get an idea of what Berkeley thinks seeing meaningfully is like. But this is a far cry from understanding what is involved in mastering a skill like hearing words meaningfully, and so Berkeley's account may seem incomplete.

ences I have had, which are tangible and kinesthetic, of the distances of things disposed about me. Thus the look of things from my earliest days has acquired a distance meaning. I have come to understand, from the way things look, how far away they are from me, information that is available to me tactually. Thus, on Berkeley's account, distance perception is entirely, as it were, within our grasp. We know, thanks to tangible and kinesthetic experiences, what it is for objects to be at distances from us. In order for us to be able to see distance, we just have to learn to correlate the way things look with our experience of distance. All that is required of the look of things for us to be able to see distance is that the looks vary reliably with changes in distance, experienced tactually. What makes Berkeley's theory possible is his realization that the tactile system and the visual system constitute two different sources of information. It is not correct to see each system as converging on the same truth, the location of an object at a distance. There is therefore no need to attribute distance information to the visual system for things to look to be at a distance. Distance information is entirely the province of the tactile system. When things look to be at a distance, the way things look is perfectly adequate as a sign for tangible experiences, even as words are signs for the thoughts they signify.

Berkeley's identification of the tactile as the source of information about the distance of objects involves him in a picture of what distance perception is like very different from that provided by the theory he sets up as his rival, the geometric theory. Incorporated into the account of the geometric theory is the view that when we see distance, we are using mind-dependent means, namely visual sensations, to reach a conclusion about something that is actually mind-independent, the location of the object in space. Seeing distance is thought of as a conclusion reached by the visual system, conceived as having, as part of its own resources, information that enables it to work out where its source of stimulation is located. But since what we see is always mind-dependent, when the visual system identifies the location of what it sees with the location of the source of stimulation, it is making an identification that is false. We can't really be said to see where the mind-independent object is. The object and its location are actually invisible. In telling this kind of story, the visual system is assumed to be operating as an independent sensory system. Tactile experiences enter into the picture, if at all, as a parallel means for arriving at the same truth, which is itself independent of any particular sensory means. Berkeley's theory undermines this account because, on his view, visual experiences are not the means by which we calculate something nonexperiential but instead signs for tangible experiences. The visual sign is not supposed to be the same as or to bear any necessary connection with the tangible experience it signifies, and the tangible experience is invisible simply because the tangible is not the visible; what we touch is not what we see.

Berkeley's account of distance perception, then, does exactly what he says it was going to do: show the manner in which distance is perceived by sight. His project is the positive one of showing how we do see distance and is not, as is so often supposed, that of showing that we don't. It has not therefore been an important part of his project to show distance exists only in the mind or that the objects we see don't exist outside the mind. The sort of enterprise commonly attributed to Berkeley, in which he tries to move inside the mind objects usually thought to be outside the mind, doesn't actually make much sense on his terms. An investigation of distance perception is an investigation of mental processes. Therefore the terms in which this investigation is carried out will be unavoidably mental, about what is in the mind. Berkeley shares with those theories he takes as his target the view that seeing is a mental process and so what we see is mind-dependent, in the mind. Where Berkeley takes issue with his rivals has to do with the way in which, on the basis of what we see, we are said to reach conclusions about the things we don't see, about things said to be mind-independent. Thus the interesting battles are fought not over the question In what sense are the things we see in the mind, mind-dependent? but instead over the question In what sense are the things we see without the mind, mind-independent? Standard interpretations of Berkeley take the part of his theory that is most controversial to be his idealism and tend to try to focus the direction of his argument around that. But, in fact, Berkeley's discussion of distance takes his idealism, such as it is, for granted. The focus of his argument is over the issue that was to become central to his immaterialism, that of the existence of that which is mind-independent.

6

Seeing Size

Berkeley's discussion of size perception parallels his account of distance perception in many respects.[1] The section on size, like that of distance, falls into two parts. The first, sections 52–78, puts forward Berkeley's own theory of size perception and hammers home its superiority to its geometric rival by showing how well it handles a well-known perceptual puzzle, in this case, the moon illusion. In the second, sections 79–87, Berkeley draws some consequences for the notion of size, relying on what has been revealed about the distinction between tangible and visible size. The clear similarities to the structure of the argument about distance make it easy to come away with the impression that Berkeley is just doing for size what he has already done for distance. Nevertheless, Berkeley's account of size perception has had a very different history from his account of distance. This part of the *New Theory* has not received anything like the attention that has been accorded some other parts of the work. Berkeley's account of size perception does not lend itself to the kind of negative interpretation that has been prevalent with respect to distance. The problem of size perception is based on quite straightforward phenomenological facts, as that what we see can take up more or less of the visual field and so can appear visually

1. The visual perception of magnitude is discussed by Pitcher, *Berkeley*, pp. 34–41, and by Armstrong, *Berkeley's Theory of Vision*, pp. 39–41.

of any size. It is, for this reason, much easier to grasp than was the "one-point" argument about distance. It has been therefore considerably less tempting to suppose that Berkeley is putting forward a claim that size is unperceivable, which can be refuted by pointing out that we just do see size.[2] Philosophers have been much less prone to discover in Berkeley's account of size the outrageousness they endeavor to refute in his account of distance. Berkeley's theory of size perception, on the other hand, has not had the success among psychologists that was the case for his theory of distance perception. Instead, the prevailing theory continues to be the one Berkeley wished to refute, that we perceive size by taking account of distance.[3]

The section on size perception deserves closer attention, however. It is true that Berkeley presents a theory of size perception that is analogous to his account of distance perception, as a process in which visual cues suggest otherwise unrelated tangible ideas. Even though Berkeley adds nothing new to the account of psychological processing developed to account for distance perception, still the picture of Berkeley as putting forward a positive program for solving some perceptual problems receives additional support. In addition, Berkeley's account has wider-ranging consequences. His understanding of the nature of size embodies serious criticisms of the account of our knowledge of extension that Malebranche had derived from his own geometric theory of size perception. So, by allowing him to concentrate on some serious issues about the nature of extension, Berkeley's theory of size perception might well be regarded from his perspective as of deeper importance than the theory of distance perception that supports it.[4]

2. Although this is a temptation into which some have fallen. This appears to be the position Samuel Bailey is prepared to argue for in his *A Review of Berkeley's Theory of Vision* (London, 1842).

3. See William Epstein, "Historical Introduction to the Constancies," in *Stability and Constancy in Visual Perception*, ed. Epstein (New York, 1977).

4. Some of the connections between Berkeley's views on size perception, microscopes, and the nature of our ideas of extension are discussed by Genevieve Brykman in "Microscopes and Philosophical Method in Berkeley," in *Berkeley: Critical and Interpretive Essays*, ed. C. M. Turbayne (Minneapolis, 1982), pp. 69–82.

The Problem with Size Perception

Much of Berkeley's discussion of size perception is even terser than his corresponding discussion of distance. Indeed, the most conspicuously enthymematic feature of his argument about size perception is the absence of an initial explanation why size perception is a problem. Since Berkeley begins with a rejection of a geometric solution to the size problem, he apparently expects his readers to be well aware of the reasons why geometric theorists think size is not one of the immediate objects of vision. It is fortunately not difficult to construct an account of the problem with which Berkeley could suppose his readers to have been familiar, an account confirmed by the sorts of issues Berkeley emphasizes in his own later conceptualization of the problem and by the somewhat richer account of the nature of the problem he provides in the *Theory of Vision . . . Vindicated and Explained*. It is necessary to keep in mind, however, that the problem Berkeley is addressing is one he shares with his geometric rivals, since his aim here, as it was for distance, is to show the superiority of his theory of vision over a geometric theory. Berkeley's own discussion of the problem of size perception occurs within the framework of the distinction between *visibilia* and *tangibilia* he has just established in the course of the discussion of distance perception. Nevertheless, the existence of the problem of size perception by sight does not arise from this distinction, from the fact that we can't see what we touch, but is instead the result of some simple anatomical and geometrical facts about vision.[5]

There is a problem with the visual perception of size because there is a discrepancy between what we see with respect to size and the visual information about size registered on the retina. As I look about the room, I see objects of specific, determinate sizes. I see the chair at the end of the room as large, certainly large enough for me to sit on, and I see the mug in front of me as quite small, small enough for me to lift easily. These size judgments don't vary as I move about the room. I continue to see the chair and the mug as

5. I am disagreeing with what Pitcher says (*Berkeley*, p. 34) in introducing his discussion of size, if he means to be saying that Berkeley's *problems* with size arise from his heterogeneity thesis. I am not disagreeing if he is instead saying that the heterogeneity thesis is important to Berkeley's *solution*.

having fixed sizes, the chair large and the mug small. But even though I see the objects as having a determined size, the way these objects are registered on the retina does not have a determined size. Projections of the objects on the retina can take up any amount of the retina at all. Moreover, I will continue to see the chair as large enough to sit on when its image takes up very little of my retina, and I will continue to see the mug as small when its image bulks very large. Thus if we want to say we see by virtue of projections on the retina,[6] we will also have to say we can't read off the size of an object from the size of what is projected on the retina. Just as the "one-point" argument tried to show what is represented on the retina is inadequate with respect to distance, so we can also argue, because of the way size is represented on the retina, that what we see is inadequate with respect to fixed size. Size, then, like distance, is not immediately perceived by sight.

It is a little easier to capture what is happening phenomenologically with respect to size than it is with respect to distance. If I hold my hand at arm's length and then move it steadily toward my eyes, I am aware of the hand I am seeing as taking up more and more of my visual field. In the *Theory of Vision Vindicated*, Berkeley tries to exhibit the notion of the size of what we see in the visual field by asking us to imagine "a diaphanous plain erected near the eye, perpendicular to the horizon, and divided into small equal squares. . . . The eye sees all the parts and objects in the horizontal plain, through certain corresponding squares of the perpendicular diaphanous plain. Those that occupy most squares have a greater visible extension, which is proportional to the squares" (*TVV* 55). We have a visual size problem because the size we experience of the object we are seeing does not reflect the size registered by the eye. A man walking toward us is seen as taking up more and more space in the visual field but is not seen as getting larger. Solving the size problem requires showing what else, what further information besides retinal size, is available to the perceiver that can disambiguate the retinal image and result in a determinate size judgment.

6. As, it will be remembered, both Descartes and Malebranche, whose geometric solutions to this problem Berkeley will want to criticize, did choose to say.

The Geometric Solution (NTV 52–53)

Berkeley's account of how we perceive size by sight is going to supplant, he tells us, "the opinion of some that we do it by angles, or by angles in conjunction with distance" (*NTV* 52). The idea that the size of an object is perceived through the mediation of its distance is widespread. Berkeley could have found such an account in Descartes's *Dioptrics*:

> As to the manner in which we see the size and shape of objects, I need not say anything in particular, inasmuch as it is all included in the manner in which we see the distance and the position of their parts. Thus, their size is estimated according to the knowledge, or the opinion, that we have of their distance, compared with the size of the images that they imprint on the back of the eye; and not absolutely by the size of these images, as is obvious enough from this: while the images may be, for example, one hundred times larger when the objects are quite close to us than when they are ten times farther away, they do not make us see the objects as one hundred times larger because of this, but as almost equal in size, at least if their distance does not deceive us. (Sixth Discourse, p. 107)

Malebranche gives a very similar account in his *Elucidation on Optics*:[7]

> But though the image on my retina of a child I see at ten feet from me is equal to the image I have of a giant thirty feet away, I yet see the giant three times larger than the child, because there occurs in my eye something that I can use to discover the difference in their distance, such as the image of intervening bodies between me and these two people, or the different disposition of my eyes when I fix them on each of them to see them distinctly, or one of the other means that can serve to disclose the distance of objects. . . . Thus, knowing the distance

7. Similar sorts of accounts can be found well before the time of either Descartes or Malebranche. For example, John Pecham in his *Perspectiva Communis*, written in about 1263, says: "However, as experience proves, the faculty that apprehends size considers the magnitude of the distance [from the observer] and not only the angle" (Proposition 74, p. 147, in *John Pecham and the Science of Optics*, ed. David Lindberg [Madison, Wisc., 1970]).

> between the child and the giant, and having besides this, according to the assumption, a perfect knowledge of optics, which teaches me that the images of objects on the retina must diminish with their distance, I give myself a perception of the giant three times the height of the child's. (sect. 27, p. 734)

We are able to see, for example, that two objects which project images of different sizes on the retina are in fact of the same size by recognizing that the objects are at different distances. That is, it is very easy to see why a retinal image of some object could be of many different sizes. As Berkeley says, "It is well known that the same extension at a near distance shall subtend a greater angle, and at a farther distance a lesser angle" (*NTV* 53). Therefore it is to be expected that light from two objects of the same size will be reflected on the retina at different-sized angles, depending on the difference in the distances, resulting in images of different sizes. What Descartes and Malebranche in the quoted passages have done is to use the same geometric fact that explains the optic fact about the size of the retinal images to explain as well a psychological fact, that objects which project different-sized images on the retina may appear to the perceiver as of the same size. A geometrically based (and mechanical) account of how the eye works is also used as the basis for an account of how perceivers see. A problem that objects whose representations on the retina are of different sizes are perceived as being of the same size is solved by attributing to the visual system knowledge of the geometric principles that account for the different sizes of the representations on the retina. The outcome of such a process, described by Malebranche as a natural judgment, one made for us by the visual system and not made by us, will be a perceived or phenomenal size. The perceived size will reflect the size the object, perceived under its optic angle, ought to be at its distance.

Berkeley's Refutation of the Geometric Theory (NTV 52–53)

Berkeley actually begins his discussion of size perception with a relatively bald dismissal of a geometric theory of size perception. His refutation falls into two parts. The first, in *NTV* 52, consists simply in a rapid reminder that the conclusions of the discussion of distance perception (that lines and angles are not a visual means for perceiving distance) must also apply with respect to size. The second part of his refutation, discussed in *NTV* 53, is more interesting. Here Berkeley directly attacks the view that the judgments we make about the perceived size of an object are a function of the distance we take the object to be. He says we fall into this mistake because retinal size does covary with distance. There are a series of cues that, as has been demonstrated, suggest distance to us, even though these cues are not necessarily connected with distance. There is nothing in their nature which requires that these cues be *distance* cues, they have merely been reliably and habitually connected with experiences of (tangible) distance. These same cues will also covary with respect to size. Since nothing about the cues per se makes them distance cues, these same cues are equally available if we are casting about for an explanation of our ability to make determinate size judgments. The "distance" cues will also be reliably and habitually connected with experiences of variations in size. The same cues, therefore, are just as much size cues as distance cues. There is no need to assume the visual system needs the geometric information that size is necessarily connected with distance in order to perceive size. The visual system is not calculating what size an object at a particular distance ought to look. The system responds to a set of cues as size cues even as it responds to these cues as distance cues: "They have as close and immediate a connexion with the magnitude as with the distance; and suggest magnitude as independently of distance as they do distance independently of magnitude" (*NTV* 53). Just as there is no need to suppose the visual system must know the size of an object before it can know its distance, so the visual system can perceive size without first perceiving distance.[8]

8. I am therefore agreeing with Pitcher that Berkeley's point is not that size has nothing to do with distance but rather that making a distance judgment is independent of and hence not causally responsible for making a size judgment. Pitcher however assumes that Berkeley thinks that cues function as size cues independently of distance because he discovers no calculations from one to the other when he introspects. This is undoubtedly so, but Berkeley's case is stronger. Given that there is nothing about a distance cue that links it necessarily with distance, it would require special argument to show that the cues we are responding to when we make size judgments are in fact distance cues on the basis of which size judgments have been

Berkeley's Conceptualization of the Size Problem (NTV 54–55)

Berkeley's procedure in his discussion of size is first to dismiss the geometric solutions to the size problem (*NTV* 52–53) and only then to set up the problem of size perception in terms of the theoretical framework he had already established in solving the distance problem. He first reminds us the distance problem was solved by distinguishing two different kinds of objects of sight, two different kinds of things we can learn about through seeing. There are visual cues, which are the immediate objects of sight, and there are the mediate objects, that which we learn about through the suggestion of these cues. Both the way in which we see and the way in which we touch involve sizes. We see greater or smaller expanses of light and color, and we touch or are aware kinesthetically of moving through greater or smaller expanses. Berkeley is making available therefore, for use in describing the size problem, a distinction between an immediate visual size and a mediate tangible size.

Before he actually sets up the size problem, however, Berkeley makes one further point about the way in which we apprehend size by sight or by touch. He says in each case, there is a minimum "beyond which sense cannot perceive" (*NTV* 54). Berkeley seems to regard this point as relatively uncontroversial. He says this is something of which "everyone's experience will inform him" (*NTV* 54). It is a claim, therefore, whose truth we are supposed to be able to establish simply by thinking about our own experiences with size perception. Berkeley would find Malebranche in agreement with him on this point. Malebranche, for example, speaking of visual perception, says: "As far as vision is concerned, a mite is only a mathematical point. It cannot be divided without being annihilated" (*ST*, bk. I, chap. 6, sect. 1, p. 27). The way in which the size of the mite is perceived by sight has a lower bound; Malebranche does not want to say that mites themselves cannot be divided. Malebranche, too, is of the impression that there are sensible minima, below which

calculated, rather than serving directly as size cues. Certainly introspection can't form the basis for such an argument. The importance of Berkeley's point about the independence of size and distance judgments has been stressed to me by Robert Schwartz.

our senses cannot perceive. Difficulties in understanding the notion of a sensible minimum have arisen as a result of misunderstandings about the role sensible minima are supposed to serve for Berkeley. These minima are a feature of a kind of psychological measuring device, intended as a way of getting at various of our psychological capacities. As such, they bear a strong resemblance to the just noticeable difference, or *jnd*, the sort of unit contemporary psychologists have introduced to measure various psychological capacities to discriminate, among them color discrimination. Berkeley's appeal to experience is not intended to be a claim we apprehend in minima. Just as contemporary psychologists interested in scaling abilities to make color discriminations do not assume the world is seen as a patchwork of just noticeably different colors, so Berkeley is not imagining the world as we experience it is composed of tiny dots. Typically, of course, we experience sizes much greater than the minimum. Minima are not to be thought of as units that compose or make up a sensory world.[9] They are not special kinds of things we apprehend but a way of talking about or measuring our capacities to apprehend, in this case, to apprehend size. The claim there is, for example, a *minimum visibile* is a claim that our capacity to experience sizes has a smallest limit or lower bound.

The reason why Berkeley is so confident there are minima lies in the nature of what it is to be apparent, or perceivable. We can't perceive without perceiving something, some content that is present to consciousness, or apparent, or noticeable. Just as we can't see without perceiving something, something noticeable, so differences between things are perceivable only in noticeable increments. It is unintelligible to suppose there can be a continuum of apparent magnitudes to zero, a continuum of magnitudes between each of which there is a lesser apparent magnitude.[10] An apparent magni-

9. This seems to be the way Armstrong is assuming that minima are supposed to function. Not surprisingly, he is unable to see how such a view has anything to do with Berkeley's topic, the difference between ideas of sight and of touch.

10. As Brook proposes. Brook's formulation seems to suppose that there is a smaller apparent magnitude than the smallest magnitude that we can see that is somehow a part of that smallest magnitude. But if it is apparent, then we can see it, and so *it* is the smallest apparent magnitude and not the larger magnitude containing it. These are the sorts of difficulties that arise if Berkeley's minima are taken to be things seen instead of ways of seeing.

tude is one that is perceivable and can be perceived only as having some determined content or extent. Since we are incapable of registering things by sight except as having a certain size, there will be a smallest discriminable size.[11]

Berkeley therefore approaches his discussion of the size problem in an order that reverses the way in which he talked about distance perception. In the case of distance, Berkeley first proposed a set of cues that could account for our ability to perceive distance by sight and only then drew attention to the consequence that the experience of distance is only immediately available to touch. He sets up the size problem, however, as one in which we must explain how visible objects of varying size can be connected with the fixed sizes available only to touch. Having called our attention to the fact that both the way in which we see objects and the way in which we touch them involve size, Berkeley's account of the problem to be solved in *NTV* 55 proceeds by identifying "the magnitude of the object which exists without the mind, and is at a distance,"[12] as one that can be apprehended tangibly but not visually, on the grounds that the visible object lacks a fixed size. Thus Berkeley supposes we will be prepared to accept that the problem of how we identify a fixed size by sight is one that, like the problem of how we identify distance by sight, is to be solved by associating visual cues with tangible ideas.

Berkeley's Solution (NTV 56–66)

The essence of Berkeley's solution was already given by him in his refutation of the claim that size is perceived through taking account of distance. Since nothing in the nature of the distance cues ties them necessarily to distance, beyond their being associated in our experience with variations in distance, then, so long as size covaries

11. See also the account of the minimum visible by Pitcher, *Berkeley*, pp. 34–35, by Brook, *Berkeley's Philosophy of Science*, pp. 67–76, and by Armstrong, *Berkeley's Theory of Vision*, pp. 42–45. For an argument that Berkeley's minima are extensionless, see David Raynor, "'Minima Sensibilia' in Berkeley and Hume," *Dialogue*, 19 (1980) 196–200.

12. This is one of the passages in which Berkeley appears explicitly to commit the "vulgar error" of assuming that what is touched is a mind-independent object.

with distance, any distance cue is equally suitable as a size cue, since it will also be habitually associated with variations in size. Nevertheless, Berkeley singles out three cues that he seems to regard as of special importance (*NTV* 56). First, he mentions that the size we see by and large varies directly with the size we touch. In general, if an object takes up a great deal of the visual field, we expect to encounter a large tangible size. It is not possible to predict tangible size simply from visible size, however, since we also make use of two other cues that vary reciprocally with visual size. If what we see is both large and confused, like our own hand held close to our eyes, we anticipate a smaller tangible size, whereas if it is both small and faint, like a distant tower, we expect a larger tangible size. Thus the visual cues that to Berkeley are the most important cues to tangible size are visual size, confusion (or distinctness), and faintness (or clarity).

The significant difference between Berkeley's account of distance perception and the account he sets up as its rival is that, according to Berkeley, our ability to perceive distances by sight does not rely on assumptions about necessary connections between the immediate and the mediate objects of sight. Instead, that we can recognize by sight how far away something is from us is to be explained as a learned ability to associate visual cues with conceptually unrelated tangible ideas of distance. It is important to Berkeley's account of size perception that he be able to make the same claim, that in recognizing by sight the size of an object we are not relying on necessary connections between the immediate perceived size and the mediate size (*NTV* 58). Berkeley's conclusions about the lack of necessary connections ultimately allow him to draw serious consequences about our abilities to make sensory based judgments about extension. Berkeley's claim that confusion and faintness are in themselves size cues is an important tool for him in his argument that size perception does not rely on necessary connections, since it is undeniable that confusion and faintness are in themselves unconnected with variations in size. The harder case to make, although the one that in the end is of greater significance, is that visual size is conceptually unrelated to the size we perceive by sight the object to be (*NTV* 59).

It is indeed an odd-sounding claim to make, that the size of an object we learn about in seeing bears no necessary connection with the visible size registered by the visual system. Nevertheless, what Berkeley wants to establish is that we pay attention to visual size information only to learn about the fixed size of the object we apprehend through touch. Our interest in how much or how little space an object is projecting on the retina or takes up in the visual field derives entirely from what we can learn about the size of the object we will feel as it impacts on our bodies. What we are interested in learning is what a tangibly sized object looks like, how to recognize from visible cues the size the object will feel to be. Berkeley argues that the phenomenon of size constancy, that an object which takes up more and more of the visual field is seen as being of the same size but approaching, has an obvious explanation. We always focus on the size we will feel as suggested by the visual cues, and since that tangible size remains unchanged, our perception of the size of the object, despite the changes in the visual field, remains unchanged. We have learned to recognize those visual changes which represent an object of a constant tangible size. We recognize the sight of a man getting larger in the visual field as representing a man who remains six feet tall. The visual size continues to change, but the tangible size represented in each change does not.

Berkeley insists a perception of the approach of a man six feet tall cannot be a perception that estimates how many visible feet high the man is, since there is no determinate visual apprehension that could be used as a measure of feet or inches (*NTV* 61). The only fixed unit that could be used as a measure of feet or inches is a unit that is experienced tangibly. I can lay a one-inch-long rigid body on top of another and establish through touch that the one completely covers the other. Even though the way we typically use things like rulers is by reading off a marking by sight, the units we are learning about by sight are tangible units. It is interesting that some of Malebranche's arguments about the difficulties in judging even relative sizes by eye in fact rely on tests that Berkeley would take to involve tangible information. For example, Malebranche writes:

> We can hardly even judge with any certainty about the relation between two bodies quite close to us; they must be picked up and held against each other for a comparison, and even then we often hesitate,

> being sure of nothing. This can be clearly seen in examining the size of coins that are almost equal; here we must place them on top of each other to see with assurance whether they correspond in size. If a line is drawn on paper and another is drawn at its end perpendicular and equal to it, they will appear roughly equal. But if the perpendicular is drawn at its middle, the perpendicular will appear perceptibly longer, and the closer to the middle it is drawn the longer it will appear. The same experiment can be performed with two straws, so that to know if they are equal, or which is longer, they must be laid one upon the other, as is ordinarily done. (*ST*, bk. I, chap. 6, sect. 3, p. 32)

Berkeley would take the kind of operations Malebranche suggests to show the unreliability of visual size judgments actually to involve tangible information. There is nothing, in fact, in what Malebranche says that would support a claim that these tangible operations are unreliable.

An account of size constancy like Malebranche's, on the other hand, proceeds on the assumption that the constant size we see must be conceptually connected with the size registered on the retina. These conceptual connections are attributed to the visual system, and hence the phenomenon of size constancy is taken to be an entirely visual phenomenon. This means that the visual system is thought to contain within it two independent steps or processes: a sensory core, which in this case is a sequence of objects of fluctuating sizes; and a visual world, here an object perceived as having a constant size. Trying to think of both of these steps as elements in our visual experience can prove puzzling. For example, is the sensory core thought to provide a visual experience of which we are conscious? This does not jibe with our sense of our experiences as of objects of constant size. But if the sensory core is thought to be unconscious, then there is no longer a reason for solving the problem by talking about natural judgments. Unconscious anatomical processes form no part of a psychological perceptual theory, an account of what the perceiver does. Nor is it really satisfying to suggest in the case of size perception that a sensory core is not available to consciousness, since it is possible to become aware of the relative extent objects are taking up in the visual field. So we are left with an account that assumes two different conscious size experiences, both visual but only one of which, the less important, reflects the functioning of the visual system.

Berkeley's account instead presents a very different picture of the experience of size and size constancy. On his account, whatever experiences are said to be visual are those attributable to the functioning of the visual system. Such experiences are phenomenologically present. When we think about the visual field, we recognize that what we see is of fluctuating size. These visual experiences can acquire tangible meanings. This claim does not require us to suppose something is going on in the visual system that necessitates some knowledge there, or that we are now seeing differently. The visual experiences themselves don't change when perceivers learn to use them as visual cues to fixed, tangible size. For all that perceivers are doing is coming to recognize what a fixed, tangible size, such as can be determined through such measures as the superposition of a rigid rod, looks like. The visual system itself has been freed, as it were, from the task of trying to work out what an object under certain geometrical circumstances is supposed to look like. We see just what we are visually equipped to see. On Berkeley's account, the problem is not to correct visual experiences but to understand them.

Once our visual task is conceived as understanding visual experience, it becomes possible for Berkeley to argue that the particular tangible size meanings we have learned to assign to visual size experiences bear no necessary connections to those visual experiences. Perceiving a fixed size by sight is a matter of learning to associate the deliverances of one sense modality with that of another modality. Therefore, as he claims, we could continue to have all the same visual experiences we are now having even if they were no longer associated with the tangible experiences with which they are now associated or, indeed, even if they were associated with no tangible experiences at all (*NTV* 64). Our visual system functions as it does independently of any other sensory modality. This means there is no right way to see any tangible size. Any visual size is equally appropriately associated with any tangible size. That we learn to read visual size cues differently when they are associated with confusion and faintness shows we are by no means restricted to taking visual sizes as a direct reflection of tangible size. The way in which we perceive size visually can't be held to be intrinsically misleading with respect to tangible size. If we can, as indeed we do, learn to associate visual sizes with tangible sizes, then we are equipped to perceive the size of objects even though whatever perceptions we have are dependent on the nature of our sensory systems.

Since we can solve the size problem by describing the process in which we perceive size by sight as one in which visual information comes to be associated with a tangible meaning, and since the visual information bears no conceptual relationship to the tangible meaning, then the perception of size takes place in exactly the same way as the perception of distance (*NTV* 65). In both cases, a problem that arises because we seem to be able to learn more through seeing than we are equipped to learn by our visual system is solved by showing that the initial visual data are not supplemented by calculations relating the retinal stimulus to other extended objects but are instead supplemented by cues that suggest data from another sensory modality, that of touch. This is the lesson Berkeley wants to derive from distinguishing between the mediate and the immediate objects of sight. Neither size nor distance perception, according to Berkeley, are to be understood as purely optical phenomena, in which the visual system tries to work out from the necessarily imperfect data registered by the visual organs of the perceiver, at what distance or size an object is. Neither the size nor the distance problem is to be solved by attempting to work out the relations between extensions that can be thought of as common to both sight and touch. Instead, they are solved by a process of associating immediately visual ideas that are distinguishable from the mediate ideas derived from touch.

The Moon Illusion (NTV 67–78)

Berkeley inserts a discussion of a notorious visual phenomenon or illusion, that the moon looks bigger on the horizon than when it has risen,[13] for the same purpose as his discussion of the Barrow illu-

13. For some further accounts of the moon illusion, see Turbayne, *The Myth of Metaphor*, pp. 185–191, Pitcher, *Berkeley*, pp. 37–41, and a very interesting article by David Berman, "Berkeley and the Moon Illusions," *Revue Internationale de la Philosophie*, 154 (1985) 215–22. There has also been much discussion of the moon illusion in the psychological literature.

sion: to demonstrate that a theory that distinguishes as his does between mediate and immediate objects of sight has greater explanatory power than a geometric theory. At first glance, the historical status of the moon illusion seems to be somewhat different from that of the Barrow illusion. The Barrow illusion was recognized by proponents of a geometrical theory as a problem for that theory, whereas geometrical theorists such as Descartes and Malebranche appear quite confident of their ability to explain the moon illusion, claiming that the illusion is the result of a natural judgment that estimates size by taking account of distance. (The idea is that the presence of various distance cues leads us to believe that the moon on the horizon is further away than it appears to be when risen, and so, even thought the moon in both cases takes up roughly the same amount of the visual field, this same visual appearance is taken to be larger in the case of the horizon moon than it is for the risen moon.)[14]

Several considerations suggest, however, that Berkeley might have regarded the moon illusion as inexplicable on geometric grounds, in addition to the perhaps rather tendentious consideration that he had already exploded the geometric account of distance. The first is that Berkeley's reference to Molyneux's article in the *Philosophical Transactions*, which presented arguments against several existing attempts to explain the illusion and was most severely critical of Descartes's account involving distance, suggests Berkeley might with some justice be under the impression that no account of the moon illusion was widely accepted.[15] A second, and perhaps more important, reason is that Berkeley actually enlarges the scope of the moon illusion beyond what had been previously discussed. In Berkeley's account, the moon illusion has two parts, the first being the widely discussed problem of the disparity between the appearance of the horizontal and the meridional moon, and the second being that "the horizontal moon doth not constantly appear

14. The facts of the matter appear to be that the retinal image is actually slightly smaller in the case of the horizon moon. Molyneux mentions this in his article, cited by Berkeley, in *Philosophical Transactions*, as does Malebranche in his discussion of the moon illusion in *ST*, bk. I, chap. 9.

15. Molyneux writes: "These Thoughts, my-thinks, are much below the accustomed Accuracy of the Noble *Des-Cartes*." *Philosophical Transactions*, 187 (p. 314).

of the same bigness, but at some times seemeth far greater than at others" (*NTV* 67). It is Berkeley's contention that an adequate account must explain both of these phenomena, and he can quite fairly claim that his own solution is the only one that can account both for the horizon moon's looking bigger than the risen moon and for the difference from one time to another in the apparent size of the horizon moon, since any account that regards distance information as important will be unable to account for differences in the perceived size of the moon in the same situation.[16]

Berkeley's discussion of the moon illusions highlights the strength of a perceptual model based on suggestion rather than calculation. His solution to the moon illusion provides supporting evidence for his claim that, in perceiving, we respond to cues directly as size cues and not as distance cues from which size is calculated. His primary proposal is that the moon is perceived as larger on those occasions when atmospheric vapors result in a fainter appearance.[17] In general, a small faint object is perceived as larger than a small clear object. Since there are more vapors on the horizon than there are surrounding a risen moon, the moon will always be fainter on the horizon and hence appear larger. And since, moreover, the amount of vapors on the horizon can increase from one time to another, this same cue also explains the variations in the perceived size of the moon on the horizon. The virtue of Berkeley's explanation, he tells us, is that he has identified a perceptible cue that, unlike the retinal image, varies with the phenomenon in question.

The appeal of a cue like faintness to Berkeley is clear.[18] Berkeley

16. As far as I know, Berkeley's formulation of the problem of the apparent size of the horizon moon is not only unprecedented but also unique. Subsequent discussions, either of the phenomenon itself or of Berkeley's account of it, fail to mention Berkeley's second aspect of the problem, although Berman is an exception here, pointing out that there are two moon illusions. He is slightly more cautious than I am about attributing both illusions to Berkeley. Pitcher does mention both parts of Berkeley's formulation of the problem, but he does not apparently consider this of importance in assessing Berkeley's claim to have solved the problem.

17. Although he does also think that even on occasions when the risen moon is faint, or as faint as the horizon moon, it will still appear smaller, because the cue is functioning in an unfamiliar context, where we are viewing distant objects by looking up.

18. Berkeley's appeal to the presence of atmospheric vapors is not itself a novelty. Mention of this has apparently been common in discussions of the moon illusion at least from the time of Ptolomy. See David Lindberg, ed., *John Pecham and the Science of*

can claim to have identified in faintness a cue that applies both in explaining why the horizon moon looks bigger than the risen moon and in explaining why the moon at the same situation on the horizon looks to be of different sizes, when an explanation involving distance fails to apply. Even though any distance cue, not just faintness, will covary with size and hence be available as a size cue, Berkeley can claim to have identified a cue that nevertheless functions as a size cue in the absence of differences in distance. Berkeley thinks, as is demonstrated in his attempt to refute Wallis's distance-based explanation of the moon illusion (*NTV* 77), that accounts involving distance are open to empirical refutation. He gives the same reason that Molyneux had given in the *Philosophical Transactions* article in rejecting Descartes's proposal, namely, that the moon on the horizon continues to look bigger when we see it in the absence of any so-called distance cues, from behind a wall or, Molyneux adds, over a smooth sea. But Berkeley also has arguments that are not in the same way empirical. An explanation in terms of distance does not have the same explanatory power as his own theory, he argues, since it fails to account for differences in the perceived size of the horizon moon. And finally, since what we see are lights and colors, not distances, what we see are as well-suited through experience to suggest size as distance.[19]

Optics (Madison, Wisc., 1970). The explanation for the importance of these vapors, however, had been geometric in nature. Increased vapors were assumed to refract differently. Malebranche, however, had argued that the refraction in question would in fact result in a smaller image, not a larger one (*ST*, bk. I, chap. 9). Berkeley is thus following his practice of showing that an explanation which had relied on a geometric context actually functions independently of that context.

19. It is important to stress this nonempirical character of Berkeley's claim, since it is sometimes said, by Pitcher, for example, that recent empirical work has refuted Berkeley. Lloyd Kaufman and Irvin Rock, for example, in "The Moon Illusion," in *Perception: Mechanisms and Models* (San Francisco, 1972), 260–68, claim to have shown that faintness is not a factor in the moon illusion. Unfortunately, they don't say exactly what they did in order to show this, since Berkeley himself says that not any kind of faintness will do, but only that which has been connected in experience with size (*NTV* 72). Berman, however, rejects Pitcher's claims on behalf on Kaufman and Rock. In any case, as Robert Schwartz has stressed to me, a defense of a distance-based model in refutation of Berkeley is not all that easy to come by since the point that is of importance to Berkeley does not depend on any particular cue, like faintness, but rather is that there is nothing in the nature of distance cues that can support a claim that they are functioning to provide distance and not size information.

Berkeley diagnoses the difficulties people have fallen into in trying to discuss the moon illusion as the result of an imperfect understanding of the distinction between mediate and immediate objects of sight. He thinks the results will be paradoxical and contradictory so long as the problem is conceived as a purely visual phenomenon. In these circumstances, he thinks, we are put in the position of trying to explain how it is "that the moon should appear of a different bigness, the visible magnitude thereof remaining still the same" (*NTV* 74). If both these phenomena are thought of visually, then we find ourselves trying to explain how it is that the moon looks bigger than it looks. We know we are equipped visually to register the moon as being of a certain size. This visual size is determined by the amount of space which light reflected from the moon takes up on the retina. We know this size doesn't alter. The moon doesn't register visually any differently, with respect to size, on the horizon or once risen. So if there is a difference in the size of the moon on the horizon and when risen, it can't be a difference that is due to vision. We explain this, Berkeley maintains, most neatly when we attribute the difference in perceived size to a different tangible estimate associated, in different circumstances, with the same visual size. Thus a distinction between the immediate objects of sight, the ones we undeniably see, and the mediate objects, suggested by the presence of the visible objects although not in themselves visible, avoids the paradoxical claim that the size we see is not the size we see. The moon illusion illustrates the difficulties that people fall into when they fail to distinguish the size we see from the size we learn about in seeing.

It is characteristic of Berkeley's account of visual size perception, as it is of his account of visual distance perception, that it does not make what Berkeley would regard as unnatural or excessive demands on the visual system itself. This is the upshot of his distinction between the immediate and the mediate objects of sight. We are held to see, strictly speaking, only what the visual system equips us to see. Whatever we are said to be able to see, to learn about in seeing, that is not an immediate object of vision derives its content for Berkeley from another sense modality, in this case that of touch. We learn to recognize by sight the circumstances under which we will undergo various tangible experiences; we learn what tangible experiences look like. These tangible experiences, although suggested by what we see, are not seen at all. There is, therefore, no discrepancy between what the visual system registers and what we end up seeing. There is no reason, then, to take the physiological processes of seeing as introducing distortions into the way what we see is registered. Berkeley's view is that so long as we don't try to attribute more or different information to the visual system than it is actually equipped to handle, we have no reason to suppose that there are inherent inadequacies in the way we see. It is this theme, that our vision is adequate and not defective, which Berkeley develops in the final paragraphs of his discussion of size perception.

7

Microscopes and the Heterogeneity of Extension

Berkeley concludes his account of size perception with nine sections (*NTV* 79–87) in which he discusses some confusions brought about by a failure to distinguish visible from tangible ideas, largely with respect to microscopes. This account of the consequences of his theory of size perception for the distinction between visible and tangible ideas, therefore, parallels the similar account of the consequences of the theory of distance perception with which Berkeley ended the distance section. His final paragraph of this section, *NTV* 87, makes the following claim:

> Upon the whole it seems that if we consider the use and end of sight, together with the present state and circumstances of our being, we shall not find any great cause to complain of any defect or imperfection in it, or easily conceive how it could be mended. With such admirable wisdom is that faculty contrived, both for the pleasure and convenience of life.

Berkeley claims, as a consequence of his theory, that we do not have reason to believe that our vision is defective. Interest in the implications of microscopes was widespread in this period.[1] The consider-

1. For example, Robert Hooke's *Micrographia* (London, 1665; rpt. Codicote, Herts., and New York, 1961) is an elaborate account of the way in which our existing faculties, particularly our vision, can be improved through the use of microscopes. Locke's *Essay* (2.23.11–13) contains a speculation about what it would be like to have microscopical eyes, which has attracted some attention. For an interesting discussion

able references in Berkeley's notebooks, the *Philosophical Commentaries*, to the problem of microscopical vision suggest, however, that the primary target of Berkeley's interest was Malebranche and, in particular, Malebranche's use of arguments concerning microscopes in his *Search after Truth* (bk. I, chap. 6), to show that we are systematically misled through vision about the nature of absolute extension.[2] The culmination of Berkeley's argument about size perception, then, is a demonstration that we are in a position to reject claims, such as that of Malebranche, that our senses are imperfectly suited to inform us about extension.

Malebranche's Argument

Malebranche's account of size perception relied on a calculation attributed to the visual system resulting in phenomenal size perception. The appeal to such a calculation received its plausibility because the calculation was thought to recapitulate the principles by

---

of the influence of the microscope on theorizing, see Catherine Wilson, "Visual Surface and Visual Symbol: The Microscope and the Occult in Early Modern Science," *Journal of the History of Ideas*, 49 (1988) 85–108.

2. The *Philosophical Commentaries* shows Berkeley playing with a number of issues concerning microscopes and size perception. For example, in *PC* 94 he writes: "Enquiry about a grand mistake in writers of Dioptrics in assigning the cause of Microscopes magnifying objects." He follows this up in *PC* 197, where he details the mistake in question and specifically refers it to Molyneux. This line of inquiry, however, doesn't show up in the published version. Relatively early on, at *PC* 116, we also find Berkeley making the following assertion: "The visible point of he who has microscopical eyes will not be greater or less than mine." This claim is pursued throughout Notebook B in entries that increasingly focus on Malebranche's claim that we can never know whether two people are perceiving the same size. Thus Berkeley seems to be exploring one line of argument in entries such as *PC* 218 and 219 which again doesn't show up in the published version. *PC* 255 and 257 include specific reference to Malebranche; the latter reads: "Malbranch out in asserting we cannot possibly know whether there are 2 men in the world that see a thing of the same bigness. v. L. 1. c. 6." In *PC* 272 and 277, we find Berkeley reaching the argument he actually used in *NTV*. For example, 277 says: "Tis impossible there should be a M.V. less than mine. if there be mine may become equal to it (because they are homogeneous) by detraction of some part or parts, but it consists not of parts. Ergo. &c." Other entries that have a Malebranchian ring concern infinite divisibility, in particular *PC* 236 and 237. The latter says: "They who knew not Glasses had not so fair a pretence for the Divisibility ad infinitum."

means of which the eye works. It is far from being the case, however, that such a calculation is held to inform the perceiver of the actual size of the object perceived. Instead, in Malebranche's hands, the nature of this calculation turns out to be a reason for arguing that our sensory judgments of size are unreliable. For one thing, the information about distance that is fed into the visual system is information available to the perceiver, that is, perceived or phenomenal distance. So we are only as good at being size perceivers as we are at being distance perceivers. And since we are only good distance perceivers at relatively short distances, whatever accuracy we might achieve as size perceivers will presumably fall within these limits.

Calculations about size, moreover, also include facts about the extent of the perceiver's own body, since one element in the calculation is the size of the retinal image. Therefore our ability to make size judgments, in addition to being inaccurate, depends upon the state of our own body and is relative to it. It is in making this claim that Malebranche relies most heavily on the evidence of microscopes. He points out that since we can, with microscopes, see previously undetectable objects or see them in greater detail than we can with the naked eye, we must recognize the world of extended objects is quite different from the world we see. With the naked eye, for example, I might see a mite as a barely detectable smudge, but with a microscope I might see it as having eight different segments. Malebranche supposes what I am seeing through the microscope is more like the way the mite or, indeed, matter in general really is, since he is prepared to say the fact that we can see more and smaller parts of more and smaller objects through a microscope is a reason for taking matter to be infinitely divisible or, perhaps, provides visual support for our rational conception of matter as infinitely divisible. Malebranche's view is that when I look at the mite with the naked eye and when I look at the mite under a microscope, the same piece of mind-independent matter bears the appropriate causal connection with what I see. What I see is a geometric transformation of what is out there, worked out on the basis of appropriate calculations. The nature of this transformation depends upon the geometry of the viewing apparatus. Some transformations, such as those depending upon the use of microscopes, result in a perception more like the object out there. The way the mite looks through the microscope, having eight segments, is more like the mind-independent mite than is the smudge seen with the naked eye. It would be a mistake, however, to conclude that through the microscope I am seeing the mite as it really is, for, if I had a better microscope, I might see the mite as having sixteen segments. The appropriate conclusion is that I never see and, no matter what instrument is devised, I never will see the mite as it really is. The way the mite is seen is altered by and dependent upon the nature of the viewing apparatus.

Even though an appreciation of the geometric connections that exist between the object seen and the way we see it demonstrates we see the object better at some times than at others, it also convinces us that we never see the object as it is in itself. The imperfections revealed by the microscope in our ordinary vision are irremediable. Malebranche says,

> Our eyes furnish us with none of the ideas of these things that we discover with microscopes or by reason. Through sight we perceive nothing smaller than a mite. Half a mite is nothing if we accept the testimony of vision. As far as vision is concerned, a mite is only a mathematical point. It cannot be divided without being annihilated. Our sight, then, does not represent extension to us as it is in itself, but only as it is in relation to our body; and because half a mite has no significant relation to our body, and can neither preserve nor destroy it, our sight hides it from us entirely. (*ST*, bk. I, chap. 6, sect. 1, pp. 27–28)

The size judgments we make express at best the size an object is relative to our bodies. We are unable to make any absolute size judgments. Malebranche illustrates this point by asking his readers to imagine that, on the one hand, God has created a miniature world, about the size of a ball, with everything on it proportionally reduced in size; or, on the other, that God has created a much bigger world than ours, with proportionally larger inhabitants. He argues that the people in the small world and the people in the vast one would make very different size judgments than we would about the objects in their worlds. A distance considered vast to an inhabitant of a tiny world would be judged to be very small by one of us. Therefore, Malebranche concludes, we should recognize that the way in which we perceive phenomenal size leaves us in no position to make claims about the absolute size of anything in the physical world.

The Molyneux Man (NTV 79)

Before embarking on his discussion of the implications of the distinction between visible and tangible ideas for microscopic vision, Berkeley draws out the consequences of such a distinction with respect to the Molyneux Man. Berkeley's claim is that again we know what the experience of the Molyneux Man would be like because we now know there are no necessary connections or any resemblance between the size we see and the fixed tangible size suggested by the size we see.[3] In the absence of any experience linking the two, the Molyneux Man would respond to visible size information without any reference to his beliefs about the tangible size of objects. The discussion of the Molyneux Man underlines the point that there is a way of expressing size judgments which is entirely a function of the way we see and is independent of any other way of taking in size information. So long as we restrict our attention to what it is like to see size or differences in size, to the immediate objects of vision, seeing size is entirely a matter of comparing portions of the visual field. What we apprehend visually is the extent or, as Berkeley says, the number of points covered in the visual field. Any time what we see almost entirely fills the visual field, then what we are seeing (the way in which we are apprehending by sight) is large, and any time what we are seeing takes up very little of the visual field, then what we see is small. If what we are seeing takes up the smallest possible extent discernible in the visual field, then what we see is as small as

3. Berkeley's intuitions about what the Molyneux Man's visual size experience would be like are shared by Marius von Senden, who writes: "Variations of apparent size are equally unintelligible [to someone newly sighted]. He cannot grasp such differences, because during his blindness he could only deal with tactual objects by direct contact, so that they have always presented the same 'sizes' to him; before operation he had no idea of 'going away from' a thing. He therefore finds it surprising that the same object can be now large and now small." *Space and Sight* (Glencoe, Ill., 1960), p. 296.

we can see, and any time what we are seeing is of a barely discernible extent, it will be as small as we can see. The Molyneux Man can't answer questions about the size of what is seen in terms of a fixed size apprehensible by touch, since these fixed sizes do not yet play a part in his visual experience. The Molyneux Man will be able to talk about his experiences strictly in terms of seeing, in terms of the size information he is aware of as he sees.

The Nature of Visual Size Perception (NTV 80–81)

The kinds of claims Berkeley wants to make about the distinction between visible and tangible ideas require, for their intelligibility, the point he has made with respect to the Molyneux Man, namely, that the Molyneux Man would, and the rest of us could, answer the question What is the size of what you are seeing? strictly in terms of the nature of the visual experience in question. By focusing on vision as a faculty that functions independently of any other faculty, Berkeley can undercut such claims as that some ways of seeing size are more like the size of the object than others or that no way of seeing size is just like the size of the object. He rests his argument on the observation that "the *minimum visibile* is exactly equal in all beings whatsoever that are endowed with the visive faculty" (*NTV* 80). Berkeley defends this claim, that the *minimum visibile* is of equal size in all beings that can see, by arguing that any talk of comparing one *minimum visibile* with another, as comparing that of a person with that of a mite, is unintelligible if it presupposes that a *minimum visibile* has parts that are not apparent to its possessor. He takes a claim that the *minimum visibile* has parts to be a consequence of saying that the *minimum visibile* of a mite is smaller than that of a person. Such a comparison requires that it would have to be possible for someone to lay the *minimum visibile* of a mite next to that of a person so as to observe that the *minimum visibile* of a mite is a part of that of the person. This possibility is absurd according to Berkeley because it supposes someone's *minimum visibile* could have unnoticed or indiscernible parts, yet a *minimum visible* is a measure of what can be noticed or discerned. Talk of *minimum visibilia* is a way of getting at the nature of the visible experience, what the viewer is aware of visually. A report about the way in which we are experiencing something is not, Berkeley says, a report about something that can "have any existence without the mind of him who sees it" (*NTV* 81). Whether or not what we see can exist mind-independently, the way what we see is being experienced by a perceiver, as large or small, cannot. A very small object can have parts smaller than can be seen, but a *minimum visible* shouldn't be confused with a very small object. Indeed, an object of almost any size (consider a star) can be in one circumstance or another minimally visible to a perceiver. A report on the part of a perceiver that something is minimally visible is a report of exactly the same experience, no matter what object is being experienced or whose visual faculty is doing the experiencing.

Berkeley is resisting the claim that a mite or a person with a microscope sees differently because the objects seen are of a smaller size. That the smallest discernible point in the visual field is the same size in a person and in a mite can seem surprising only if the size of the object seen, its physical size, is confused with the way its size is experienced, its phenomenal size—or if there is assumed to be a connection between the two. It is certainly true there are objects whose size is smaller than the size of any of the objects I can see, and it is probably true mites see objects of a smaller size than any I can see. It is also true if I become able to see these very small objects through a microscope, their physical size remains unchanged. But, according to Berkeley, it is not true, as Malebranche's argument supposes, that because we are now seeing objects that were previously too small to see, our ability to see sizes, to discriminate sizes visually, has been altered. Imagine I am gazing across a field, and I am able to make out a faint smudge of barely discernible size. I advance upon it (or it strolls toward me), and my visual field is progressively filled with something cat-shaped, until I get so close to the cat I can see a mite, now barely discernible to me, roosting on the cat. Even though initially the object that was barely discernible was a cat and subsequently it was a much smaller mite, my ability to make visual discriminations hasn't improved or changed one iota, although it is of course perfectly true my visual experience has been changing. This, I take it, is Berkeley's point when he writes: "No exquisite formation of the eye, no peculiar sharpness of sight, can make it [the *minimum visibile*] less in one creature than in another"

(*NTV* 80). The ability to make size discriminations is the same, no matter what the size of the objects being looked at or how close the viewer is to the object. The *minimum visibile* is a unit measuring visual experiences that can be expressed independently of any way of measuring the object seen. Questions about the size of a *minimum visibile* are questions about visual capacities, not questions about the size of the object being seen.

Visual Defects and the Microscope (NTV 82–86)

Berkeley has established, then, there can be no differences whatsoever in the experiencing of something as minimally visible. There can be no sense, therefore, in which one visual faculty can be considered defective as compared to another with respect to its capacity to experience the minimally visible. He proceeds to analyze the nature of visual experience in order to discover in what sense a visual faculty might be considered defective. Our visual experience, after all, is rarely confined to the minimally visible, to a single visible point, but generally takes in the entire visual field. Berkeley claims, with respect to the visual field, there can be no differences in the size of the visual field of a perceiver (*NTV* 82). Just as we can't see less at some times than at others, so also we can't see more. On those occasions, Berkeley says, when we speak of ourselves as seeing more, as when we cease to confine our attention to a small room and look out over some vast terrain, we haven't increased the size of the visual field. The "more" we speak of seeing does not represent an increase in the number of visual points in the visual field but refers to a greater number of tangible objects suggested by the visual scene. Berkeley's claim is that "of these visible points we see at all times an equal number" (*NTV* 82).[4]

It is possible, however, to imagine a visual faculty whose field contained more points. If we ask how our vision might be improved, one way might be to have a bigger visual field. A second way in

4. Robert Schwartz has pointed out to me that Berkeley's claim is not quite true, since it is possible that the size of the visual field be *reduced*, through damage or just by closing one eye. There is no obvious way, however, of extending the visual field.

which vision might be improved has to do with the fact that what we see on the edges of the visual field is confused and out-of-focus. A more perfect visual faculty would enable us to see more of the visual field in focus. We can certainly think of possible improvements in the visual faculty, but they must be such as to actually improve the way we see (*NTV* 84). Berkeley, therefore, is able to argue that microscopes do not in the same way constitute an improvement in the way we see. They neither enlarge the visual field not improve its clarity. We continue to see through a microscope in the same way we do without it. We do, however, see a different visual display with the microscope than we do without it. As Berkeley says: "A microscope brings us, as it were, into a new world: It presents us with a new scene of visible objects quite different from what we behold with the naked eye" (*NTV* 85). Even though the way we see through a microscope is not different, what we see is different.

Nevertheless, even though what we see through the microscope is different from what we see with the naked eye, it is not the case that what is seen is better as a way of seeing size. Consider again the example of seeing a cat advancing over a field. From the time the cat is seen as a smudge until it is at arm's length, the cat is seen differently from moment to moment. It is not the case, however, that some one among these many ways the cat is seen is the way the cat's size looks. This is because when I see the size of the cat, I am seeing what a fixed, tangible size looks like. So whether I am looking at the smudge across the field or at a cat at arm's length, what I am seeing is just as much what the tangible size looks like in the first case as in the second. It is true the second case is more likely to be fully informative of what the tangible size is, so there is a sense in which the viewer is likely to have a better idea of the cat's size when the cat is viewed at arm's length. But what makes one way the cat looks a better indicator of size than another has to do entirely with the ease with which tangible size is suggested. The size that is seen in a successful seeing of the size is tangible size, not some one privileged form of visual size. For this reason, Berkeley is inclined to say that the different view we see through a microscope is not in fact a better view because what the object looks like through a microscope is in general not correlatable with tangible size.

There is nothing in Berkeley's argument, it should be noted, that would undermine the enthusiasm of someone like Robert Hooke for the use of the microscope. It is true Hooke refers to the microscope as a way of perfecting our vision, but he means that, with the microscope, there will be other sights to see, not that the microscope reveals imperfections in our ordinary vision. Hooke instead, speaking of what the microscope shows in terms not dissimilar to Berkeley's, regards the microscope as a way of extending the usefulness of observation. Hooke writes: "By the help of the *Microscope*, there is nothing so small, as to escape our inquiry; hence there is a new visible world discovered to the understanding" (*Micrographia*, Preface). It is no part of Berkeley's program to deny that what we see through a microscope looks different from what we see with the naked eye. His concern is only to deny that from what we see through a microscope, we are in a better position to calculate how things are. Berkeley's view, too, is in the end not all that different from what he would have found in reading Locke. For example, Locke writes:

> Nay, if that most instructive of our Senses, Seeing, were in any Man 1000, or 100000 times more acute than it is now by the best Microscope, things several millions of times less than the smallest Object of his sight now, would then be visible to his naked Eyes, and so he would come nearer the Discovery of the Texture and Motion of the minute Parts of corporeal things; and in many of them, probably get *Ideas* of their internal Constitutions: But then he would be in a quite different World from other People: Nothing would appear the same to him, and others: The visible *Ideas* of every thing would be different. (2.23.12)

We find Locke, like Berkeley, emphasizing that what is different with a microscope is our "visible ideas" and drawing the conclusion that such a difference, in isolation, is not very helpful.

The Adequacy of Visual Size Judgments

Berkeley's interest in microscopes derives from their use in an argument that we are systematically misled by vision about size or extension, such as the one made by Malebranche. From Berkeley's point of view, we could be misled only if we were led astray by vision about tangible size. The sorts of considerations Malebranche brings forward are not sufficient to show that vision is misleading, since

Malebranche's arguments suppose that we are seeing a size common to both sight and touch. Consider again Malebranche's example of the miniature and the huge worlds. Malebranche's claim was that we can never know what is the absolute size of what we are seeing, since we must admit the people on the tiny or on the vast world would make very different perceptual size judgments than we would about the absolute size of objects on their respective worlds. Such an argument presupposes, however, that there is no way to correct visual size judgments with tangible information. Suppose we are equipped on our world with a ruler, with the sort of rigid instrument that, according to Berkeley, can be used to establish fixed size judgments through touch. We can imagine ourselves stepping to the vast world, still equipped with our ruler, and continuing to make size judgments as before, by laying objects on the ruler and ascertaining how many units are covered from one end of the object to the other. There is no reason to suppose that a big person on this world wouldn't concur with our judgment, since the fact that a big person has bigger hands than ours wouldn't interfere in any obvious way with the measuring process. The possibility that these different-sized worlds exist does not constitute an argument that size is imperceivable. It is quite true that if one of us and one of them look at the same ruler from the same distance, what they are seeing will take up less space in their visual field than what we are seeing will take up in ours. Visual size judgments will indeed differ. So long as making size judgments, however, is understood to be a matter of connecting a tangible size with an unrelated visual size, there will be no reason to suppose one of the two people will be unable to see the size of the ruler. Learning to see fixed sizes on the larger world is merely a matter of recalibration.

Berkeley's account of size perception, like his account of distance perception, is an account of how we do see size. He is trying to show that seeing size, like seeing distance, is not a matter of calculation on the part of the visual system, in which the visual system attempts to work out, from the visual size, what the absolute or mind-independent size of the object ought to be. Instead, Berkeley's claim is that visual size does not bear some sort of calculatable relationship with absolute size but is instead a visible cue by means of which we recognize what a tangible size looks like. As in the case of distance,

Berkeley can argue that what we need in order to perceive size is within our sensory grasp. It is true the fixed or determined sizes that we see are nonvisible. But it is not true that the nature of sensation and its reliance on corporeal processes leave us uncertain about the nonvisible nature of the extended world. The sizes we see are nonvisible simply because they are tangible.

At the end of the section on distance, Berkeley established the basic principle of the heterogeneity of sight and touch. He claims to have shown that a visual sight is not the same as and does not resemble a tangible feel. The way things look is entirely different from the way they feel. At the end of the section on size, Berkeley has sought to establish a consequence of this distinction. The way in which we perceive size by sight has been thought to be unavoidably inaccurate, because it is unavoidably perceiver-dependent. A size perception derived from retinal size can reflect only size from the perspective of the perceiver and not absolute size. Visual size perception will always be judged to be imperfect if we think of the visual system as trying and failing to calculate the absolute size of the object perceived. This picture of the visual system as defective is undercut, Berkeley claims, when we recognize the independence of the visual and tangible sensory systems. What are taken to be the deficiencies of the visual system with respect to absolute size are not shared by the tangible system. We have sensory access to the fixed size of objects, but such access is tangible not visible. It is absurd, moreover, to suppose the visual system provides an imperfect representation of what we touch, since the way we see is different from the way we touch. To the extent, therefore, that visual size information can be correlated with tangible size information, the sizes things look to be provide perfectly appropriate cues to the sizes that can be felt. Visible extension is not a sight-dependent version of extension-in-general but what is perceived by an independent sensory system. And, in general, neither the ways things look nor the ways things feel are versions of something else; they are instead the expression of the way the visual and tactual systems work. Berkeley's discussion of size perception provides support for a picture of how the various sensory modalities relate to one another. They don't replicate one another but instead can be seen to supplement one another by providing entirely different sorts of information.

8

Perceiving Situation by Sight

The third problem that Berkeley promised to solve was to show the manner in which the mind perceives by sight the situation of objects. Berkeley structures his account of the visual perception of situation entirely differently from his accounts of distance and size perception, so that, even though he covers most of the same points in all three sections, these points occur in the discussion of situation perception in a different order. This difference in organization reflects, for one thing, the fact that Berkeley is further along in his own argument. The distinction between ideas of sight and ideas of touch, which Berkeley had to develop out of his demonstration of the nature of distance perception, is by now well in hand. Berkeley is able to use the device of the Molyneux Man this time as an aid to conceptualize what are the different ideas of situation available to sight and touch. Thus the discussion of the Molyneux Man occurs as a part of the formulation of the problem of situation perception instead of forming a part of a discussion of the consequences of a solution to the problem, as is the case in Berkeley's account of distance and size perception. The most striking difference in the way the section on situation perception is organized, however, has to do with the prominence given to the problem of the inverted retinal image. The problem of the inverted retinal image, like the Barrow illusion and the moon illusion, is a problem that Berkeley thinks is mishandled by the geometric account but can readily be dissolved by his own theory. Although he uses the first two problems in the preceding sections as a kind of pivot in his argument, hammering home his own account before turning to talk about its consequences, the discussion of the problem of the inverted retinal image appears instead as a frame to Berkeley's theory of situation perception. He raises the problem of the inverted retinal image before giving any account at all of situation perception and concludes the section with his solution to the problem. This has the result of giving a great deal of importance to the problem of the inverted retinal image. Indeed, it is not at all uncommon for commentators to ignore the discussion of situation perception entirely and to announce that the subject of Berkeley's third section is the inverted retinal image.[1] This overconcentration on the inverted retinal image is, I think, a mistake. It not only encourages the reader to overlook what Berkeley explicitly calls the subject of the section, how we perceive the situation of objects by sight, but, as well, without a thorough consideration of the material on situation perception makes it much harder to see what it is Berkeley found problematic about the inverted retinal image.

The Geometric Theory and the Inverted Retinal Image (NTV 88–89)

One reason, perhaps, why Berkeley found it necessary to introduce the problem of the inverted retinal image into his initial presentation of the nature of situation perception is that the inverted retinal image is an integral part of the geometric account of situation perception and dictates some of the peculiarities of this account. Berkeley thus raises the problem of situation perception in terms of what everyone knows about the retina. He says: "There is at this day no one ignorant that the pictures of external objects are painted on the retina, or fund of the eye: That we can see nothing which is not so painted: And that, according as the picture is more distinct or confused, so also is the perception we have of the object: But then in this explication of vision there occurs one mighty difficulty. The

1. Armstrong, for example, says: "We pass on to the third section of the *Essay* (sections 88–120), which is devoted to the problem of the inverted retinal image" (*Berkeley's Theory of Vision*, p. 45). Warnock and Pitcher take a similar position.

objects are painted in an inverted order on the bottom of the eye" (*NTV* 88). There is a problem about how we perceive the situation of objects in space because of what Berkeley presents as a well-known fact: a perceiver can see nothing which is not registered on the retina, and the way the perceiver sees is determined by the nature of what is recorded on the retina. The reason why a nearsighted person sees most objects as blurry is because light from the object is focused behind the retina, giving rise to blur or confusion on the retina itself. But light from an object also focuses on the retina in an inverse order, that is, light from the upper part of the object focuses on the retina below light from the lower part of the object. So if the way what is registered on the retina determines the way we see, then surely, just as a blurry image, from a presumably distinct object, is seen as blurry, so an inverted image ought to be seen as upside down.

According to this picture, perceiving the spatial position of objects must be a two-stage process, because the information recorded on the retina does not by itself account for what is perceived. In this case, the problem is that the retinal information, though geometrically related to what is perceived, is nevertheless not the same as what is perceived. Additional information must be ascribed to the visual system in order to explain the transition from what the perceiver's retina records to what the perceiver actually sees. The geometric account assumes we perceive the position of objects in space as we do their size and distance, by recapitulating the causal history that created the retinal image. The mind works back from the point where the light hits the retina to the point in space from which the light originates, and by these means it constructs a phenomenal representation seen to be in the same spatial location and orientation as the physical object which is its cause. Here, for example, is Descartes's account of the process:

> As to position, that is to say the direction in which each part of the object lies with respect to our body, we perceive this with our eyes in the same way as we would with our hands; and this knowledge does not depend on any image, nor on any action which proceeds from the object, but only on the position of the small points of the brain whence the nerves originate. For this position, changing ever so little each time that of the members where the nerves are inserted changes, is estab-

> lished by nature not only in order that the mind may know how each part of the body which it animates is placed with respect to all the others, but also so that it may transfer its attention from there to any of the locations contained in the straight lines that we can imagine to be drawn from the extremity of each of these parts, and prolonged to infinity. In the same way, when the blind man of whom we have already spoken above moves his hands A towards E, or similarly C toward E, the nerves inserted in that hand cause a certain change in his brain, which gives his mind the means to know, not only the location A or C, but also all the others which are in the straight line AE or CE, so that [the mind] can turn its attention to the objects B and D, and determine their locations; yet for this he does not need to know, or to consider at all, the location of his two hands. And similarly, when our eye or our head turns in some particular direction, our mind is informed of this by the change which the nerves inserted in the muscles used for these movements cause in our brain. In the eye RST, for example, it is necessary to assume that the position of the small fiber of the optic nerve, which is at point R, or S, or T, corresponds to a certain other position of the part of the brain 7, or 8, or 9, which enables the mind to know all the locations along the line RV, or SX, or TY. So that you must not be surprised that the objects can be seen in their true position, even though the picture they imprint upon the eye is inverted: for this is just like our blind man's being able to sense the object B, which is to his right, by means of his left hand, and the object D, which is to his left, by means of his right hand at one and the same time. (*Dioptrics*, Sixth Discourse, pp. 104–5)[2]

According to Descartes's account, we see a phenomenal representation derived from retinal information, the orientation of which is calculated by tracing the light rays, from the angle at which they hit the retina, back to the object. Even though the retinal image itself is

2. Descartes illustrates this account with a little picture of the blind man with his sticks, which Berkeley reprinted in an Appendix to the second edition. Malebranche does not mention the fact of inversion at all but, in his Elucidations on Optics, gives a similar sort of account, based on the fact that the retinal image is also reversed from right to left. "I know that it [the horse I am seeing] is standing with its head turned to the right, although its image is reversed on my retina. For since I know that my retina is not flat but concave, geometry teaches me that perpendicular lines on a concave surface necessarily cross and that they can be parallel to each other only when they fall on a plane surface, and that therefore I must judge that its situation is opposite to that of its image" (*ST*, Eluc. on Optics, p. 745).

inverted, what we see, based on this image, is not. It might be worth pointing out that, on the geometric account of situation perception, the only information about situation that presumably is assumed to be actually in consciousness is that provided by the reinverted phenomenal representation.

Seeing the Retinal Image (NTV 88–89)

Before a geometric solution to the problem of situation perception can be accepted as a serious solution, it is first necessary to accept the problem of the inverted retinal image as a serious problem. For situation perception is presented as a two-stage process, requiring geometric calculations, because we can see only what is painted on the retina. Berkeley's claim is that his own account of situation perception provides a better solution to the problem of the inverted retinal image than does the geometric account. But then Berkeley too must see the problem of the inverted retinal image as serious, one that deserves a solution. But if Berkeley takes the problem of the inverted retinal image seriously, then this in turn suggests he too thinks we see only what is painted on the retina. It seems, then, that Berkeley would have to be prepared to endorse what he says everyone knows: "That the pictures of external objects are painted on the retina, or fund of the eye: That we can see nothing which is not so painted" (*NTV* 88). There is no way, however, that Berkeley could endorse this particular claim consistent with the position he develops in the course of his argument, for the end of the section finds him explicitly denying "that the pictures of external objects are painted on the bottom of the eye" (*NTV* 117). It seems as though there is no way Berkeley could take the inverted retinal image seriously unless he were prepared to agree that the immediate objects of vision are little pictures painted on the retina. But the position he actually endorses is that what appears on the retina is tangible and not visible, and hence the retinal image can in no sense be among the immediate objects of sight, since it is not seen at all. So the question that arises is why Berkeley would suppose there is a problem with the inverted retinal image, since the retinal image is neither seen nor in consequence seen to be inverted with respect to what is seen. Berkeley would seem to have a much easier way of dissolving the problem at his disposal: since we are not in visual perception aware of anything inverted, the inverted retinal image is nothing more than an artifact of the geometric theory, which, like lines and angles, actually plays no role in how we see.[3]

Since it is consistent with Berkeley's final position to hold that retinal images are not the immediate objects of sight, that what we see is not what is painted on the retina, and since the contrary view, that Berkeley thought that what we see is what is painted on the retina, is very hard to reconcile with the rest of his theory,[4] several commentators have held that Berkeley never intended to accept the position on which the retinal image problem is based but instead was willing to adopt it provisionally or hypothetically because he was interested in the conclusions that could be drawn from it.[5] But it is not entirely satisfactory to say that Berkeley was willing, in this case, to mention facts about what appears on the retina solely for the sake of argument. For, as D. M. Armstrong points out, Berkeley includes facts about the retina in other arguments as well, most notably in the "one-point" argument for the nonimmediacy of distance perception.[6] In fact, in general, it seems as though when Berkeley takes over the problems of space perception from geometric theorists such as Descartes and Malebranche, he takes over from them a view about the importance of the retina in determining what we can see. Berkeley's ability to identify the immediate objects of vision, what we can see simply by virtue of having a visual system, seems to rest for

3. That the problem that is usually taken to be the subject of Berkeley's third section ought not to be a problem for him at all is raised by numerous commentators, including Armstrong, Warnock, and Pitcher.

4. And is, moreover, contradicted by one of the *PC* entries, which reads: "Mem: To discuss copiously how and why we do not see the Pictures" (*PC* 268).

5. This is the position that Turbayne takes in "Berkeley and Molyneux on Retinal Images." Pitcher also seems attracted to a view rather like this one.

6. Armstrong makes this point both in his book and in a reply to Turbayne, "Discussion: Berkeley's New Theory of Vision," *Journal of the History of Ideas*, 17 (1956) 127–29. Armstrong, presumably because he thinks Berkeley has no right to include facts about the retina in his theory of vision, expresses this point rather more negatively than I have done. He thinks Berkeley got trapped, in his account of distance perception, into identifying the points on the retina with the immediate objects of vision and therefore was "hoist with his own petard," unable to repudiate this claim, when he turned to the problem of the inverted retinal image.

him, as for Descartes and Malebranche, on the fact that it is the retina that is photosensitive, so that what we see and the way we see depend upon the way in which light is reflected on the retina. For this reason Armstrong and Gary Thrane have argued that Berkeley did in fact identify the immediate objects of vision with the retinal image, although Thrane, who sees this position in a more positive light than does Armstrong, is careful to point out that this does not mean Berkeley thought we see what is painted on the retina as if with a third eye; rather, the immediate objects of vision, light and colors, are the ways in which what is recorded on the retina reaches visual consciousness.[7]

The kinds of arguments Berkeley gives, particularly in discussing the nature of the immediate objects of vision, make it difficult to adopt one sort of extreme position, namely, that the retinal image is an artifact of the geometric theory and hence irrelevant to how we see. On the other hand, it is also difficult to accept, at least without further analysis, Armstrong's position that Berkeley would have identified the immediate objects of vision with the pattern on light on the retina, since Berkeley also seems to have been inclined to assert, and also to have set considerable store by the assertion, that the pattern of light on the retina is tangible. And if the pattern of light on the retina is tangible, then one thing it can't be, in Berkeley's theory, is a proper object of vision. The right answer to the question What are the proper objects of vision? is always, for Berkeley, light and color.

We don't, however, see just any light and color. Those lights and colors which we see at any given time are only those that have been projected on the retina at that time. It would seem, then, more appropriate to take Berkeley's actual position to be that the proper objects of vision, patterns of light and color, correspond to tangible patterns of light on the retina. This way of putting things allows Berkeley to recognize the importance of light rays on the retina in determining what we see while it also denies that events on the retina are themselves visual phenomena. It will not be the case for Berkeley that there are two kinds of things to be seen, or two kinds of

7. Gary Thrane, "Berkeley's 'Proper Object of Vision,'" *Journal of the History of Ideas*, 38 (1977) 243–60.

things that can be registered visually, a retinal picture and a phenomenal representation. On this view, Berkeley is not identifying what we see with an inverted image on the retina; he is asserting the relevance of facts about the retinal image to what we see.

Alan Donagan, although he finds this an attractive position, with some textual support, thinks it is one that Berkeley, if he adopted it, actually had no right to.[8] He sees it as in conflict with Berkeley's characterization of the properly "philosophical" theory of vision in *TVV* 43, which is to be limited to "how the mind or soul of man simply sees" and must be distinguished from a geometric account of particles or an anatomical account of the "mechanism of the eye." Therefore, Donagan thinks, the only facts that ought to be available to Berkeley in characterizing, for example, the immediate objects of vision, are phenomenological ones, reports by perceivers about how they see. I think, however, Donagan is interpreting Berkeley's methodological strictures a little too narrowly and that Berkeley did not intend to rule out as irrelevant to a study of vision either facts about the anatomy of the eye or facts about the geometric relations between the eye and the object we are seeing. Berkeley's strictures properly concern a matter of levels of inquiry. He thinks we shouldn't deploy facts that exist at one level, the level of optics or anatomy, to solve problems that exist at another level, the psychological level where we ask about how the mind sees as it does. He thinks the geometric theory has involved itself in such a confusion of levels when it tries to explain what perceivers do with the information received by eye by attributing to these perceivers or to their visual system a recapitulation of the causal account of how visual information reaches the eye.

Berkeley's rejection of this use of the causal history of visual information, however, in no way commits him to a rejection of the causal history itself or to a rejection of the optical and anatomical information involved in this history. His claim is simply that a causal history is insufficient to account for how we see and must be supplemented by a psychological account, which, as a theory that explains what perceivers do with the information they register, is a different theory from one that explains how the information got registered in

8. Donagan, "Berkeley's Theory of the Immediate Objects of Vision."

the first place. There would be no reason for Berkeley to deny the relevance for a theory of vision of facts such as that seeing requires a photosensitive retina and that a perceiver will see only such objects as reflect light on the retina. He does of course want to maintain that such facts are tangible facts, and hence whatever correlations that can be established between such facts and facts about what we see will be contingent and based on experience.[9] Therefore, that the pattern of light on the retina is inverted will not be, for Berkeley, irrelevant to a theory of vision, although, of course, according to Berkeley, any account of this inversion must involve the fact that what is inverted is tangible.

Berkeley's Rejection of the Geometric Theory (NTV 90)

Berkeley rejects the geometric theory of situation perception on exactly the same grounds that he rejected the geometric theory of distance perception: because the theory assumes an unjustified use of lines and angles by the perceiver in coming to perceive the spatial situation of objects. He says: "To me it seems evident that crossing and tracing of the rays is never thought on by children, idiots, or in truth by any other, save only those who have applied themselves to the study of optics. And for the mind to judge of the situation of objects by those things without perceiving them, or to perceive them without knowing it, is equally beyond my comprehension" (*NTV* 90). The geometric theory incorporates reference to lines and angles into its account of how we see, without providing the only sort of evidence that would be appropriate to the presence of these lines and angles, namely, that the perceiver is aware of them.

Again, at first glance, it might seem Berkeley's criticism is unfair. The geometric theorists are describing unconscious processes allegedly carried out by the visual system, for perceivers but not by them. But again, Berkeley's criticism is actually intended to raise doubts about the viability of the description of the process that is supposed to be carried out by the visual system, inasmuch as label-

9. In *TVV* 51, Berkeley makes the point that such experientially based correlations can be established through the use of the imagination, even though we never actually feel the impact of light on the retina.

ing the process unconscious does not absolve the theorists from all responsibility for justifying the particular content of the process that is assumed to be involved. Berkeley is calling attention to the fact that, as a piece of psychology, the geometric theory contains some unsupported features. After all, the geometric theory does not maintain we just see objects in their normal orientation. It says that what we see is a phenomenal representation whose spatial orientation is the result of specific calculations. The calculations by means of which the phenomenal representation is constructed are such that the orientation of the representation is isomorphic with that of the object. The geometric theory provides a picture of vision as a two-stage process, in which what we see is the result of specifiable calculations from an initial presentation. For this picture of vision as a two-stage process involving calculations to be viable, there must be good reasons for identifying a first stage in the visual process. Unless there were a reason for saying there had to be a first stage in vision of a particular sort, there would be no grounds for supposing there must be a mediate calculation stage, nor would there be any evidence to allow us to fill in the gap with a calculation of a particular sort. According to the geometric theory, we don't perceive objects directly, because we perceive the results of these objects' interactions, through the mediation of light rays, with the photosensitive retina. Therefore the kinds of facts about the retina of which Berkeley says in *NTV* 88 that no one is ignorant are what allow us to describe the first stage in visual perception.[10] Berkeley's case is that the geometric theorists, in order to justify the particular calculations they ascribe to the visual system, in fact end up investing the retinal image with more information than can be supported by any available evidence. Among the first-stage information they are forced to include specific information about the angles at which the light rays hit the retina and the lines that can be traced back from these angles. We could say information about light rays and angles are registered on the retina only if perceivers were visually aware of lines or angles, just as we know the retina registers light and color, because perceivers are visually aware of light and color. But perceivers are not similarly aware of lines or angles. Therefore there is no evidence

10. Although Berkeley will later argue that calling what is on the retina a picture confuses the issue.

that we perceive situation by means of calculations of lines based on angles.

Why There Is a Problem with Situation Perception (NTV 91–97)

Berkeley's project is to show he has a better way of solving the problem of situation perception than that provided by the geometric theory, one that can do a much better job of explaining the orientation of what we see. But before he can set about this project, he first has to reconceptualize the problem of situation perception on his own terms. He has to show there is a problem with situation perception because the immediate objects of vision are not the same as the mediate objects properly belonging to touch. This is not an easy or straightforward task. The problem of situation perception for geometric theorists derives entirely from the unexpected fact of the inversion of the retinal image, which shows that information about spatial orientation can be registered on the retina in a manner that does not correspond to the orientation of what is perceived. Without this esoteric fact about the retina, it would seem we might never have guessed there was a problem with situation perception. For if we just think about how things look, we might suppose that the visual field itself has a spatial orientation, that what we see presents itself as up, down, right, and left, just as it presents itself as colored, and that we just register this orientation. In the same way that we just see what colors objects are, so we just see where they are. What Berkeley intends to show is that the picture of what vision is like on which this supposition is based, in which we see an organized visual field, is a faulty picture. His demonstration of this is going to be to show that the terms by means of which we pick out spatial orientation, like up, down, left, and right, have a meaning that is limited to the tangible. Such terms can't be, except through the mediation of tangible ideas, applied to the visual field. Therefore there is a problem with situation perception because situation, in itself and immediately, can't be perceived by sight.[11]

11. I am claiming that Berkeley's actual position with respect to situation is the same strong position that he held for distance, not only that situation can't be

Berkeley makes use of his thought experiment about the Molyneux Man to disentangle the tangible meanings of situation terms from the visual signs that they typically accompany in the experience of the sighted. He first develops an account of the ideas of situation as they are perceivable by touch. His technique here is to concentrate on the experience of someone who is still blind and to show that such a person would have an immediate apprehension by touch of situations like up and down. Berkeley describes this blind man's experience as follows:

> By the motion of his hand he might discern the situation of any tangible object placed within his reach. That part on which he felt himself supported, or towards which he perceived his body to gravitate, he would term lower, and the contrary to this upper; and accordingly denominate whatsoever objects he touched. (*NTV* 93)

An object is identified as being in a spatial position by the blind man when he makes a movement of reaching out and touching it. By means of this movement, the object is perceived to be in a specific location with respect to the perceiver. These movements are feelings of tending in a direction. In addition to the direction out from the perceiver, such movements include a feeling of weightiness, of being pulled toward the earth. Berkeley's claim is that we call the direction in which we feel ourselves pulled, 'down,' and we call the opposite direction, 'up.' We identify that part of the object we touch when we reach in the direction of gravitational pull, 'down,' or 'lower,' and we call the part we touch when we reach against gravitational pull, 'up,' or 'higher.' So what Berkeley can claim to have shown is that direc-

---

immediately perceived by sight but that, in its own nature, it is not the sort of thing that can be seen. If this is right, then it provides additional reason for rejecting the position that Berkeley thought that what is immediately perceived by sight is a two-dimensionally ordered array. Evidence, therefore, against my interpretation is that in *TVV* 44, Berkeley does describe the visual array as "various in their order and situation," although it is possible that the variety Berkeley has in mind here is William James's "buzzing, blooming confusion" rather than an organized visual field. Evidence in favor of my interpretation, apart from the general account I shall give, is Berkeley's insistence in *NTV* that terms like 'up' and 'down' apply to the visual field only analogically as they would, for example, to music.

tionality is built into any sense of movement out in space and is therefore one of the immediate objects of touch.[12]

The next point Berkeley makes is that this sense of directionality is limited to what can be perceived through such movements. If 'down,' for example, refers to the experience of being pulled in the direction of gravity, then it cannot have any application in circumstances where there can be no such experience of a weighty or downtending movement. It can apply only to things toward which we can make movements of reaching out and touching. It is only under those circumstances that a perceiver is confronted with "such as were conceived to exist without his mind in the ambient space" (*NTV* 94). It is clear that no thought, desire, or passion can be reached out for or touched. They cannot be apprehended through any sort of movements out from the body. Therefore, when we speak of ourselves as being in the heights of ecstasy or the depths of despair, we intend to apply these terms metaphorically. Since we have no experience of making movements toward such emotions, we are not led to locate them in the space through which we make our movements or to invest their locations with an outness or mind-independence. Instead, we take these emotions to be, as indeed they are, mind-dependent ways of feeling, to which directionality terms cannot literally apply. If I understand the trend of Berkeley's argument, he is saying that whenever we have the experience of reaching out in a direction to touch, we can conceive what is touched to have a mind-independent location, but in the absence of such movements we never assume a mind-independent location. He is not saying that only things that exist outside the mind can be touched, nothing that

12. Pitcher, in his discussion of situation perception, writes: "Tangible objects, in the system of the *Essay*, exist around us in real physical space. There is no difficulty in understanding what terms such as 'up,' 'down,' 'erect,' 'top,' 'bottom,' 'high,' 'low,' and so on, mean when applied to objects of that sort" (*Berkeley*, p. 43). But this, it seems to me, even apart from the implied equation of 'phenomenal' and 'unreal' space, ignores Berkeley's attempt to show what such terms mean to a perceiver who applies them kinesthetically. No matter where the objects are, the blind man is still faced with the problem of working out how to tell when an object is high and when it is low. He solves this problem in the manner Berkeley describes, by reaching out and touching, and thus comes by his ideas of high and low etc. So I think Pitcher exaggerates the importance of the claim that tangible objects are "in space" and not "in the mind," although this certainly is a feature of Berkeley's argument.

exists in the mind can be touched. His argument does not, therefore, hinge on a supposition that tangible objects exist "outside the mind."

In the next section, *NTV* 95, Berkeley appeals to the Molyneux Man in order to show that visual experiences do not, in their own nature, provide any information about the situation of what we see. In his argument, Berkeley refers back to section 41, part of the discussion of distance perception. His idea seems to be that since, as has already been shown, what we immediately see contains no distance information, it can contain no situation information either. This is presumably because, as he has just shown, the situation of an object is immediately perceived through the kinesthetic experience of reaching out and touching. But we reach out and touch only those things which are located at some distance. Since, as has been shown, the Molyneux Man would not be able to read any distance information from his first experience of an array of light and color, he would have no reason to take what he is now seeing to be something toward which he could reach out and touch, and so he would not take what he is seeing to be the sort of thing to which situation terms are applicable. What he is aware of visually would seem to him to be like other nonkinesthetic ways of apprehending; they would not be the sort of things toward which we can make movements. There would be nothing about the presence of some color or colors that could suggest to the Molyneux Man that an object would be touched if he reached in some direction or other. There is nothing about being aware of colors, then, that suggests situation. The Molyneux Man would come to change his mind about what he sees only because what we see can be reliably correlated with and hence used as signs for things we reach out and touch. It is only under these circumstances that we can take what we see to have a mind-independent location. We can generalize from the case of the Molyneux Man, then, to the claim that there is nothing about seeing per se, but only seeing that has been correlated with touch, which can convey situation information.

The Molyneux Man, therefore, would not initially be in a position to transfer his kinesthetically based understanding of 'up' and 'down' to what he sees. Nor would he, Berkeley argues, be able to apply more complex orientation terms, such as 'erect' or 'inverse,' terms that describe the orientation of complexly organized objects in space (*NTV* 96). The Molyneux Man when blind would have a tangibly based understanding of such orientation terms. This means he has complex tangible ideas of, in Berkeley's example, parts of the human body. He has an organized sense of what a head feels like, what a hand feels like, and what a foot feels like. He also has a complex idea of what the earth feels like, toward which he perceives things tending and which he therefore calls 'down' The Molyneux Man would call an object 'erect' according to its orientation with respect to the earth. If he runs his hands, starting from the earth, up a person's body, then he will call that person's orientation erect if he encounters the feet first, next to the earth, and the head last, furthest from the earth. But on the occasion of the Molyneux Man's first being made to see, he would not take what he is seeing to be in some orientation, "for he never having known those terms applied to any other save tangible things, or which existed in the space without him, and what he sees neither being tangible not perceived as existing without, he could not know that in propriety of language they were applicable to it" (*NTV* 96). Orientation terms are applied by the blind man to tangibly organized objects to which the tangibly based definition of 'up' and 'down' can be applied. But there is nothing in what the Molyneux Man initially sees that would reflect this tangible organization, so there would be nothing in what he sees that would suggest to him that orientation terms would apply.

Berkeley's Solution to the Situation Problem (NTV 97–100)

There is a problem with the way in which we perceive situation by sight, because, although situation is properly and immediately perceived by touch, what we immediately see does not in itself provide situation information. But we clearly can perceive situation by sight, for we can tell, just by looking, where to reach in order to touch some object. Therefore situation is a mediate object of sight, inasmuch as what we see has come to mean the presence of specific, tangible locations. Since we are able to reach for what we see, we have learned how to recognize what tangible locations look like, or, more precisely, we have learned how to correlate the way a particular movement of reaching in some direction feels with some way the visual array looks (*NTV* 97). In order for this kind of correlation between the way things look and the way reaching feels to occur, there must be present in the act of seeing some cues that can serve as a signal for some particular act of reaching, just as, for distance perception by sight to be possible, there had to be cues that accompany vision, such as accommodation and convergence, which could serve as signs for the presence of a long or short distance. Berkeley's proposal is that it is the experience of the way the muscles of the eye feel when we turn our eyes up that suggests we are seeing 'up' and hence can serve as cues for when we must reach up (*NTV* 98). We learn to reach up for the objects we have raised our eyes to look at, just as we have learned to anticipate a journey of a particular length when we experience accommodation and convergence. Berkeley, in talking about muscle movements, has again fastened on the same cue that Descartes mentioned in his account of what it is to look up. Again, the important difference is that Berkeley, unlike Descartes, does not suppose the way the muscles move functions as a cue because of any geometric relationship in which the muscle movements stand to the object that is looked at. In order for a cue to contingently suggest some movement or other, it need do no more than covary with the movement in question. The experience of feeling the eyes move up must reliably covary with the feeling of reaching an object by moving up.

Berkeley does claim, however, that there is a correlation between what we see when we turn our eyes up and what is registered on the lower portion of the retina. Berkeley thinks we deserve an explanation for why we say we look up at those objects from which light is reflected on the lower part of the retina. He says: "And this seems to me the true reason why he should think those objects uppermost that are painted on the lower part of his eye: For by turning the eye up they shall be distinctly seen; as likewise those that are painted on the highest part of the eye shall be distinctly seen by turning the eye down, and for that reason esteemed lowest" (*NTV* 98). Optically speaking, it is a fact that what is recorded on the bottom of the retina comes into focus when we raise our eyes up. One part of Berkeley's solution involves the assertion of facts about the retina vis-à-vis what we see. The only thing he appears to want to deny is that visually speaking, what we see when we turn our eyes up so as to bring the lower part of the retina into focus is literally up or can be thought of as the upper part of the visual field. This is because the immediate objects of sight don't have a situation. ("For we have shewn that to the immediate objects of sight considered in themselves, he would not attribute the terms high and low" [*NTV* 98].) Thus we are not correlating what is on the bottom of the retina with something that is visually up, for nothing can be visually up. We apply the term 'up' merely analogically to what we see when we literally turn our eyes up, which is also when we bring the bottom of the retina into focus.

The rest of Berkeley's solution involves the claim that, with respect to vision, 'up' and 'down' literally apply to the muscles' movements involved in bringing some parts of what we see into focus and have no literal application to what we see. It is not the case that what we see when we look at that portion of the visual field which is correlated with what is reflected on the bottom of the retina is itself literally up. It is the case that we see what reflects light on the bottom of the retina by raising our eyes up and hence can refer metaphorically to that portion of the visual field as up. The means by which we identify some portion of the visual field as up, namely, upward eye movements, have a straightforward connection with bringing the bottom of the retina into focus. Such a connection, however, does not assume that we register two situations, one in the visual field and one on the retina, and correlate the two. Berkeley is not concerned to deny any of the optical or anatomical facts about vision, but he does want to deny a particular and common way of assessing what we see, in which we take what we see to be laid out before us in an array, which presents itself as up, down, to the left, and to the right.[13]

13. In *TVV* 47, Berkeley stresses that it is a mistake, but one to which we are prone, to suppose that situation terms can apply to what we see in the same sense that they apply to head movements. He says: "But, in the case of vision we do not rest in a supposed analogy between different and heterogeneous natures. We suppose an identity of nature, or one and the same object common to both senses. And this mistake we are led into; forasmuch as the various motions of the head, upward and downward, to the right and to the left, being attended with a diversity in the visible ideas, it cometh to pass that those motions and situations of the head, which in truth are tangible, do confer their own attributes and appellations on visible ideas wherewith they are connected, and which by that means come to be termed high and low, right and left, and to be marked by other names betokening the modes of position; which, antecedently to such experienced connexion, would not have been attributed to them, at least not in the primary and literal sense."

Berkeley has, then, provided an account of situation perception that deals with the issue which had concerned geometric theorists: that we identify as 'up' portions of the visual field from which light strikes the lower part of the retina. This fact about the retina, although accounted for on Berkeley's theory, is not, as it is for the geometric theory, what serves as the basis for any problems about situation perception. It is not because the retinal image is inverted that Berkeley thinks there is a problem with the way we see situation. Berkeley has therefore provided a solution not just to the problem of the inverted image but to what is actually a new problem, one that has not been previously recognized. The new problem is to explain how it is we are able to perceive situation by sight at all, given that situation is not among the proper objects of sight, is not the sort of thing that can be seen, but is available to us only through touch. In what remains of the section on situation perception, Berkeley explores the implications of his new way of conceptualizing the problem of situation perception.

9

The Nature of Visual Experience

That situation is not the sort of thing that can be seen is another among the Berkeleian contentions that appears to fly in the face of common sense. The claim, however, that it is Berkeley's concern to refute is again that of the geometric theory, in this case, that what is seen has itself a spatial organization that is isomorphic with what is represented. In the rest of the section on situation perception, Berkeley does two things. He develops an account of the nature of visual experience, by means of which he intends to make plausible the view that what we see should not be thought of as organized representations, like pictures, and he uses this account to give an explanation of what is going wrong in the way the problem of the inverted retinal image is usually conceived.

The Molyneux Man Again (NTV 101–110)

Berkeley has claimed that since situation terms have no immediate application to visual experience, a Molyneux Man would not take anything in what he first saw to be up or down, or consequently erect or inverted. In sections 101–112 he discusses a possible objection to this thesis. The objection is that

> a man, for instance, being thought erect when his feet are next the earth, and inverted when his head is next the earth, it doth hence

> follow that by the meer act of vision, without any experience or altering the situation of the eye, we should have determined whether he were erect or inverted: For both the earth it self, and the limbs of the man who stands thereon, being equally perceived by sight, one cannot choose seeing what part of the man is nearest the earth, and what part farthest from it, *i.e.* whether he be erect or inverted. (*NTV* 101)

The intuition behind this objection is clear and initially quite compelling. What the Molyneux Man would find laid out before him on first being made to see would be something like an array of visual objects. It seems reasonable to suppose that he could, therefore, just see that the head in what he is seeing is pictured as further away from the earth than are the feet. This objection presupposes that the visual field can be thought of as a kind of picture of what's out there, organized in such a way that we can just read off the spatial positions of the objects in the picture. In arguing against this objection, then, Berkeley is going to be rejecting the notion that what we see or the deliverance of vision could be taken to be a kind of picture or set of spatially organized representations.

Berkeley's argument is that the Molyneux Man's visual experience would not present itself initially as organized in any manner he could recognize as corresponding to the organization of his tangible experience. "That which I see is only variety of light and colours. That which I feel is hard or soft, hot or cold, rough or smooth. What similitude, what connexion have those ideas with these?" (*NTV* 103). Because the Molyneux Man's visual field contains within it something red and shiny is clearly no reason at all for him to suppose he is in the presence of something that will feel and taste like an apple. The array of light and colors alone can't serve initially as signs for the presence of tangibly organized objects with which he is already familiar. Nor is it the case that what he is seeing will present spatial characteristics that can be connected with the spatial properties of tangible objects. The presence of a color patch of a particular size can represent an object of any tangible size, so there can be nothing in the sizes of what the Molyneux Man sees that could suggest to him the size of any tangible object. And since "figure is the termination of magnitude" (*NTV* 105), the figure or shape of what the Molyneux Man sees can't be connected with some one tangible shape. Therefore Berkeley concludes: "In the first act of vision no idea entering by the eye would have a perceivable connexion with the ideas to which the names earth, man, head, foot, etc., were annexed in the understanding of a person blind from his birth; so as in any sort to introduce them into his mind, or make themselves be called by the same names, and reputed the same things with him, as afterwards they come to be" (*NTV* 106). The Molyneux Man's first visual experience, then, can't be taken to be a picture of anything he now could recognize; what he is now seeing is an array of lights and colors which is in no sense a representation of his tangible notions of things.

In his consideration of a remaining difficulty, Berkeley questions the supposition that the visual field could be organized in any way like a picture to the Molyneux Man, even one that the Molyneux Man couldn't at first read. The difficulty is that

> though it be granted that neither the colour, size, nor figure of the visible feet have any necessary connexion with the ideas that compose the tangible feet, so as to bring them at first sight into my mind, or make me in danger of confounding them before I had been used to, and for some time experienced their connexion: yet thus much seems undeniable, namely, that the number of the visible feet being the same with that of the tangible feet, I may from hence without any experience of sight reasonably conclude that they represent or are connected with the feet rather than the head. (*NTV* 107)

According to this objection, just because the Molyneux Man will not be able to recognize that one blotch of color will feel like a head, and a different blotch of color will feel like a foot, ought not prevent him from seeing there is only one of one kind of blotch and two of another, so allowing him to reason which blotch must represent the head and which the feet.

The objection presupposes the visual array is presented to the Molyneux Man as a set of blotches, sufficiently differentiated one from another that he can count each different blotch. One portion of his visual field contains the head blotch, even though he can't recognize it as such. The visual field is still being thought of as a kind of spatially organized picture, but one that from the point of view of the Molyneux Man is not yet representational. The flaw in this conception, according to Berkeley, is that it continues to attribute to the visual field an entirely unjustified structure or organization, assumed to be isomorphic with the tangible. Berkeley's claim is that the visual field initially lacks organization. An array of different colors will be considered part of one bundle, to be called a head, only when it has been correlated with other batches of tangible experiences that are united in one collection under one label, that of head. Coming to understand one's visual experience is not a matter of correlating a natural unit, the visual head, with a tangible head. Visual experience doesn't get organized so as to be the experience of a head except by virtue of its correlation with tangible and other experiences. The mistake Berkeley is warning against is that of supposing that our sensory experience comes organized into natural units or that number is something "really existing in things themselves" (*NTV* 109).[1] Rather, Berkeley says, number

> is intirely the creature of the mind, considering either an idea by it self, or any combination of ideas to which it gives one name, and so makes it pass for an unit. According as the mind variously combines its ideas the unit varies: and as the unit, so the number, which is only a collection of units, doth also vary. We call a window one, a chimney one, and yet a house in which there are many windows and many chimneys hath an equal right to be called one, and many houses go to the making of one city. In these and the like instances it is evident the unit constantly relates to the particular draughts the mind makes of its ideas, to which it affixes names, and wherein it includes more or less as best suits its own ends and purposes. Whatever, therefore, the mind considers as one, that is an unit. (*NTV* 109)

Whether a set of visual ideas is taken to belong to a larger collection is entirely arbitrary, therefore, since complex ideas are framed not by nature but by perceivers, according to whatever strategies or standards are convenient and useful to their purposes. We cannot

1. Berkeley considers this point to be a rejection of the claim that number is a primary quality. Number appears on various lists of primary qualities, including Locke's list in *Essay* 2.8.9. Malebranche, for example, in his Elucidation X, lists number, along with extension and essences, as those things the mind sees through a clear idea (*ST*, Eluc. X, p. 621).

take the visual field to have its own organization independent of any structure of complex ideas.

Berkeley's account of the functioning of the visual system is thus quite different from the kind of job assigned to vision by a geometric theory. On Berkeley's account, vision does nothing more than register light and color. Lights and colors are not in and of themselves representations of anything. What we see acquires a representational structure and hence can stand for something or mean something only when it is combined into collections of complex ideas by the work of the mind. Berkeley is presenting this theory in opposition to the geometric account, which conceives of the task of the visual system as one of constructing pictures, that is to say, spatially organized representations, whose spatial properties can be linked with the external world the pictures represent. In casting doubt on the notion that what we see is organized like a picture, Berkeley is undercutting the model of vision of the geometric theory.

The Heterogeneity of Sight and Touch Again (NTV 111–112)

On Berkeley's account, it is misleading to think perceiving is primarily a task that requires coordinating or developing connections between such deliverances of vision as a visible man, considered as a collection of visible ideas representing a man, and something else, a tangible or externally existing man. His claim is that through sight, we learn about those aspects of things which are visible, that is to say, those qualities of which we can be visually conscious, whereas, through touch, we learn about an entirely different set of qualities, those we are aware of tangibly or kinesthetically. Berkeley concludes, "From what hath been premised it is plain the objects of sight and touch make, if I may so say, two sets of ideas which are widely different from each other" (*NTV* 111). Since what we are aware of visually is one thing and what we are aware of tangibly is another, it is not possible to compare the spatial properties of what we see and what we touch, so as to say that they are of the same or different sizes, at the same or different distances, or in the same or different locations. These kinds of judgments can be made only in terms of comparable units, but a way of seeing doesn't have a distance or a location with respect to a way of feeling. You can see, as it were, the size of a gap between two visible points, because you can see the number of visible points between the two, just as you can feel how far distant one tangible object is from another. But neither of these two means of perceiving distance is available to compute a distance between a visible and a tangible point. There is no unit of measurement that will allow you to get from a point that is visible to one that is tangible, "the distance between them doth neither consist of points perceivable by sight nor by touch, *i.e.* it is utterly inconceivable" (*NTV* 112).

Berkeley puts his point in this way, which speaks of visible distances, in order to be able to spell out his claim that "the proper objects of sight are at no distance, neither near nor far, from any tangible thing" (*NTV* 112). It is perhaps more accurate to say that his point is that the appropriate unit of measurement for spatial concepts like distance and situation is tangible, so that whenever we talk of perceiving the distance or the situation of something we are seeing, we are not speaking of a visible distance or situation but are instead actually referring to a tangible distance or situation that has come to be associated with something in its own right uninformative with respect to distance or situation, namely, what we see. But what Berkeley is specifically concerned to deny at this point is that it be thought possible to compute a distance between something that we see and something else that is tangible or external. We cannot, therefore, think of seeing the distance of an object as projecting what we see to a distance at which the cause of what we see is located. Berkeley's claim is that the model of perception on which the geometric theory is based requires the perceiver to carry out an impossible task: to compute relations between items that are incommensurable.

Berkeley's Solution to the Problem of the Inverted Retinal Image (NTV 113–120)

Berkeley has already laid out what his own approach to the inverted retinal image is like.[2] He has previously asserted (in *NTV* 98)

2. For further discussion of the problem of the inverted retinal image, see the very complete account in Turbayne, *The Myth of Metaphor*, pp. 192–202, and Pitcher,

that there is a correlation between what we see when we look up and what is registered on the bottom of the retina. He says, moreover, in *TVV* 50, that he doesn't deny the tangible images on the retina are indeed inverted with respect to the tangible objects from which the tangible light rays go forth to hit the retina. But of course none of these tangible events on the retina is relevant to the way in which we learn about the orientation of objects by touch. What Berkeley wants to spell out is the reason why this inversion is not, as the geometric theory takes it to be, a problem for vision. He wants to show there is no problem that can be coherently expressed which requires the eye to construct an erect appearance in order to correct an inversion it is supposed to see or register immediately. There is no stage in the visual process where the eye or the mind has a motive for turning what is being seen upside down in order to make it right.

Berkeley's contention is that the problem the geometric theory sets out to solve, how it is that we see an erect appearance although the retinal image is inverted, presupposes what he has already shown to be impossible, a comparison between the spatial properties of a visual picture and those of something nonvisual. In *NTV* 113, Berkeley explains the problem as follows:

> Herein lies the difficulty, which vanishes if we express the thing more clearly and free from ambiguity, thus: How comes it that to the eye the visible head which is nearest the tangible earth seems farthest from the earth, and the visible feet, which are farthest from the tangible earth, seem nearest the earth? The question being thus proposed, who sees not the difficulty is founded on a supposition that the eye, or visive faculty, or rather the soul by means thereof, should judge of the situation of visible objects with reference to their distance from the tangible earth?

The problem the geometric theory sets out to solve is to explain how it is the man we see is erect, or is in the same orientation as the

---

Berkeley, pp. 44–48, Armstrong, *Berkeley's Theory of Vision*, pp. 45–52, and Warnock, *Berkeley*, pp. 28–31. See as well, Turbayne, "Berkeley and Molyneux on Retinal Images," Armstrong, "Discussion: Berkeley's New Theory of Vision," E. J. Furlong, "Berkeley and the 'Knot about Inverted Images,'" *Australasian Journal of Philosophy*, 41 (1963) 306–16, Thrane, "Berkeley's 'Proper Object of Vision,'" and Donagan, "Berkeley's Theory of the Immediate Objects of Vision."

tangible man with respect to the tangible earth, when what is on the retina is inverted with respect to tangible man and earth. For the problem of the erect appearance of the inverted retinal image to be a real problem, the geometric theorist has to take the eye (or rather the mind) to be working out the visual orientation of what it sees with respect to something nonvisual, the orientation of the tangible earth. The eye has to be assumed to be initially puzzled because what it immediately sees is not in the same orientation as the nonvisual world that is represented, and hence it has to be assumed to want to work out a correct orientation of what it sees. But, of course, in Berkeley's view, the eye has no information about the situation of the tangible earth and can make situation or orientation estimates only with respect to what it sees. The eye, then, never has a reason to take what it sees to be inverted since, with respect to the visual orientation, what it sees is perfectly coherent. There is no way to give the eye a motive for reinverting what it sees which does not require an illicit comparison between visual and tangible orientation.

Berkeley spells out this claim in the following two sections. In *NTV* 114, he asks us to confine our attention simply to "the proper objects of sight," and his claim is that, with respect to the proper objects of sight alone, there is no information that would lead us to take what we are seeing to be inverted. This paragraph is not without difficulty, since it has not been entirely clear what Berkeley has in mind by "the proper objects of sight." He begins by saying that when we consider the proper objects of sight, we find "the head is painted farthest from and the feet nearest to, the visible earth; and so they appear to be." Perhaps the most straightforward way of reading what Berkeley is talking about at this point would be to take him to be referring to the visual field, at such a time as we have learned to read it. Then he is simply pointing out that things appear to us in exactly the orientation in which we find them laid out before us. As he says, "What is there strange or unaccountable in this?" He then moves on to make a suggestion. "Let us suppose," he says, "the pictures in the fund of the eye to be the immediate objects of sight." This suggestion on Berkeley's part is generally taken as a proposal, possibly for the sake of argument or possibly because Berkeley thinks it is true, that we grant that the proper objects of vision, the visual field, are in fact identical with pictures painted on the retina.

It strikes me, however, that Berkeley might here be making a more modest proposal. He might simply be asking us to suppose what we are looking at are pictures painted on a retina. We are not imagining something about our ordinary visual field but instead are imagining we are doing something we ordinarily don't do, filling our visual field with the pictures painted on some retina. What appears on the retina under this supposition are pictures, things that are colored, because there are perceivers who are imagined to be looking at the retina.

What Berkeley is then pointing out is that, again, we will be seeing things in the orientation in which they are painted. He says: "The head which is seen seems farthest from the earth which is seen; and the feet which are seen seem nearest to the earth, which is seen; and just so they are painted" (*NTV* 114). There is again no discrepancy between the appearance, the ways things seem, and the way things are painted. If I am reading Berkeley correctly, what he is doing is asking us first, to consider the experience of our visual field, and second, to imagine that we are looking at a picture on a retina; and he is pointing out that, in both cases, there is no appearance/reality discrepancy, visually speaking. But, as he immediately moves on to consider in *NTV* 115, this will not satisfy the geometric theorist, who will want to point out that what we are looking at in the two instances are not in the same orientation. It is because the picture on the retina is inverted that there is a problem with the erect appearance. Berkeley's response is to argue that the picture on the retina can be judged to be inverted only with respect to something tangible. What is on the retina is internally perfectly consistent, the head is furthest from the earth, and so on, and so there is nothing that should make us want to turn it upside down other than its alleged but of course nonexistent relationship with a tangible earth.

Unless there are clues in what is immediately registered on the retina that suggest that what is so registered is in an inverted orientation, the geometric solution to the problem of the inverted retinal image is unmotivated. But it might seem the problem still remains, because one might still want to say that the retinal image we register is inverted with respect to the erect appearance we see. Berkeley's claim is that the tendency to make this response in fact reveals the mistaken supposition on which the problem of the inverted retinal image is based (*NTV* 116). The eye, or possibly the mind, can be thought to have a problem only because it is being treated as if it were in the position of someone who is looking at another person's retina. The eye is being conceived as if it were in the position of the notorious third eye that can peer at its own retina. If one person could look at another person's retina, that first person would indeed see a little picture on the other person's retina, and, as well, the first person could see the scene at which the second person is looking. So from the perspective of the first person, there would be a little picture on the second person's retina, which would be a replica of what the second person is looking at but which would be upside down with respect to the larger scene. But it is only the first person (or the third eye), who is in a position to look at both a retina and a visual scene, who can say that there is something inverted. Only a third eye, which can look at both the retina and the visual scene, is having two visual experiences whose situations can be compared with each other. Without the third eye, and going simply by what can be seen painted on the retina, there is nothing that can be judged to be inverted.

This model of the third eye, moreover, has encouraged a thoroughly mistaken view of the relationship between what is on the retina and the external world so represented (*NTV* 117). Berkeley had introduced the problem of the inverted retinal image, it will be recalled, by saying that everyone knows that pictures of external objects are painted on the retina. But, in fact, this description is antithetical to what everyone knows about the nature of external objects. External objects are not the sort of thing that can be pictured. A picture is something visual, it is made up of light and color. The experience of a picture, therefore, requires the presence of something with a visual system. We could imagine a picture painted on the retina only if we imagined someone looking at the retina. Since the pictures we experience are mind-dependent, they cannot resemble or copy an external object, which, according to Berkeley, has only tangible properties. Berkeley puts the point in terms of his own characterization of external objects, but it clearly applies to the views of geometric theorists such as Descartes and Malebranche, who are prepared to agree with Berkeley that visual qualities are ways of experiencing the world and hence don't copy or resemble properties of the mind-independent external world. A third eye does indeed experience a picture on a retina that copies a larger visual scene. But that larger visual scene cannot be identified with the external world, although Berkeley thinks this is clearly the mistake people have fallen into. The larger visual scene and the little picture on the retina are both experiences of the person or the third eye imagined to be looking at the retina and the world that surrounds it. What the little picture that is being experienced on the retina copies is another visual experience, that of looking at the world. What is experienced on the retina is not a copy of a mind-independent or a tangible world.

So the problem of the inverted retinal image rests on a series of mistakes. Even if there were a picture on the retina, and we had a third eye to see it, that picture would not be an inverted copy of the external world. A picture on the retina, moreover, in and of itself would not be judged to be inverted, and thus the claim that the retinal picture is inverted depends upon an illicit comparison between something visual and something nonvisual. But finally, in fact, what is on the retina is not a picture at all, because it is not visual in nature, except in extraordinary circumstances, when the retina is being looked at by a third eye. "Rightly to conceive this point," Berkeley says, "we must carefully distinguish between the ideas of sight and touch, between the visible and tangible eye; for certainly on the tangible eye nothing either is or seems to be painted" (*NTV* 119). Berkeley, that is, for the sake of the discussion of the problem of the inverted retinal image, has asked us to imagine that we are looking at a retina and thus seeing a picture there in order to show that such a picture could not be judged to be inverted. But he does not claim that such a picture actually plays a role in ordinary visual experience, for, in fact, what is on the retina that is operative in seeing is tangible and hence does not constitute a picture at all. Berkeley spells out this point more clearly in *TVV* 50, where he says,

> The pictures, so called, being formed by the radious pencils, after their above-mentioned crossing and refraction, are not so truly pictures as images, or figures, or projections, tangible figures projected by tangible rays on a tangible retina, which are so far from being the proper objects of sight that they are not at all perceived thereby, being by nature altogether of the tangible kind, and apprehended by the imagi-

> nation alone, when we suppose them actually taken in by the eye. These tangible images on the retina have some resemblance unto the tangible objects from which the rays go forth; and in respect of those objects I grant they are inverted. But then I deny that they are, or can be, the proper immediate objects of sight.

In *TVV* 51, Berkeley distinguishes between "pictures," which are the proper objects of sight, and "images," which appear on the retina and are tangible. But even though the distinction does not emerge in these terms in the *New Theory*, it is nevertheless clear from *NTV* 119 that Berkeley thinks what is effectively on the retina when we see is tangible and so, though correlatable in the way he describes in *NTV* 97 with what we see, is by no means identical with what we see.

The Inverted Image and Geometric Theory

Colin Turbayne, in his very interesting article "Berkeley and Molyneux on Retinal Images," has stressed the importance of the conclusion that Berkeley's discussion of the inverted retinal image has allowed him to reach, namely, that what we see cannot be considered a copy of an external, unseen world, for the only copies or originals of what we see are other things we see. Turbayne points out that this principle, established in the *New Theory*, bears an important resemblance to the Likeness Principle, that an idea can only resemble another idea, which plays a central role in Berkeley's argument in *Principles of Human Knowledge*. Turbayne also thinks that in arguing that what we see can't copy an external world, Berkeley has seriously undermined the views of such philosophers as Locke, Malebranche, and Descartes, who, according to Turbayne, held that "external objects can be seen only mediately or indirectly by means of certain immediate objects of sight which do not exist outside the mind and which are images or resemblances of external things" (p. 339). I think this view is exaggerated, since none of the men Turbayne mentions held such a crude resemblance theory of perception. Descartes, it will be recalled, took himself to be arguing against a simulacrum theory and, like Berkeley, stressed the extent to which the events on the retina must be understood to be tangible, rather than pictures. Descartes did, however, appear to fall into the trap Berkeley mentioned, that of taking what is pictured on the retina to resemble external objects, inasmuch as he begins the Fifth Discourse of the *Dioptrics* by saying, "Thus you can clearly see that in order to perceive, the mind need not contemplate any images resembling the things that it senses. But this makes it no less true that the objects we look at do imprint very perfect images on the back of our eyes" (p. 91). So Berkeley has certainly pointed to an incoherence in Descartes's treatment of inverted images, even if he has not, in this argument, exploded his entire program. Descartes's theory of vision, moreover, while not a resemblance theory, is a representational theory. What I see is not identical with but represents the external world, with which it shares a common spatial organization. Berkeley's criticism of the geometric account of situation perception speaks accurately to this aspect of Descartes's theory, since it casts doubt on the coherence of a calculation that moves from properties of what we see to properties of something unseen.

Berkeley holds, like the geometric theorists, that there is a problem with situation perception by sight, that our ability to reach for objects just by looking deserves explanation. His account of the nature of this problem, however, is very different from that embedded in the geometric theory. According to Berkeley, spatial position and orientation are not among the immediate deliverances of sight. We have an entirely satisfactory way of identifying spatial organization, the kinesthetic experience of reaching out and touching, but such experiences have no immediate application to the array of light and color we see. We must therefore learn to correlate light and color with reaching experiences so that we can come to read what we see as standing for a spatial organization determined by touch. The cornerstone of Berkeley's argument is that spatial position is entirely a matter of touch and has no immediate application to what we see. The geometric theory according to Berkeley is based on a mistaken attribution of spatial position to what is visual and what is nonvisual alike. The problem of the inverted retinal image, which serves as evidence for the geometric theory that situation perception by sight is a two-stage process, illustrates the confusions that are engendered by assuming there is to visual objects a spatial organization that must be isomorphic with the nonvisual spatial organization which such visual objects represent. These confusions are dissolved when we recognize vision does not present us with a spatially organized picture representing objects out there in space but merely shows us those qualities apprehensible by sight. The spatial organization of these qualities, like their function as representations, is not a part of what is given visually but results from our ability to form these various qualities we perceive into collections, correlated with tangible ideas of spatial properties.

10

The Heterogeneity of Ideas of Sight and Touch

In Berkeley's final section of the *New Theory* (121–159), he says he is going to look into the differences between ideas of sight and ideas of touch and to investigate whether there is any idea common to both senses.[1] The exact status of this section and the exact nature of the problem with which it is concerned is, however, a little confusing. It is true Berkeley is proceeding exactly according to the plan he announced at the beginning of the work. There he said he would first show how we see distance, size, and situation and then turn to the question whether there is any idea common to sight and touch. Now, he says, having finished the first three projects, he will turn to the last. But it is also true that Berkeley has not been neglecting the issue of the differences between ideas of sight and touch all along; instead he has included a discussion within each of the first three sections about the implications of his account for the heterogeneity of ideas of sight and touch. The relationship between these previous discussions of heterogeneity and the new section is, at least initially, a little unclear. Is the final section a summary of the preceding material, or does it instead develop new material?

1. The heterogeneity of the ideas of sight and touch is discussed by Pitcher, *Berkeley*, pp. 48–58, and by Armstrong, *Berkeley's Theory of Vision*, pp. 52–58.

The Heterogeneity of Sight and Touch (*NTV* 121–138)

The Problem with Heterogeneity (NTV 121)

In the paragraph introducing this final section, *NTV* 121, after saying that he is going to examine whether there are any ideas common to sight and touch, Berkeley sets the stage by claiming that from what he has previously shown, "it is plain there is no one self same numerical extension perceived both by sight and touch." This suggests that his earlier discussions of the differences between sight and touch have been limited to questions of numerical identity, that all that he has shown so far is that an idea of sight is not numerically the same as an idea of touch. But this in and of itself is not much of a thesis. It is relatively straightforward to grant that seeing is not touching and hence that the particular result of exercising the one faculty is not numerically the same as a particular result of exercising the other. It is hard not to see much of the earlier discussions of the relationship between ideas of sight and touch as directed to a stronger thesis than this.[2] Moreover, when Berkeley gives his reasons in the final section for supposing the ideas of sight and touch to be of different kinds, he says, quite explicitly and quite accurately (*NTV* 127), that he is drawing on arguments that have already been established. So he must think he has already shown that ideas of sight and touch are specifically as well as numerically distinct, that an idea of sight and one of touch are not just two different ideas but also two different kinds of ideas. It seems the final section on heterogeneity is, then, to some extent, a summing up of material that has already been developed rather than a presentation of new arguments.

Berkeley has certainly discussed the issue of heterogeneity with respect to his problem areas, distance, size, and situation. What he is now asking is whether there are any ideas *at all* common to sight and touch, and in answering that question he focuses on the traditional common sensibles, extension, figure, and motion, such as he would have found on lists of Descartes or Locke. Berkeley can also, then, be

2. Consider *NTV* 111 from the discussion of situation perception: "From what hath been premised it is plain the objects of sight and touch make, if I may so say, two sets of ideas which are widely different from each other."

seen to be starting off this portion of his inquiry afresh, first by considering extension and by noting that, from what he has said about the perception of distance, size, and situation, it has been established that a visual idea of extension is not numerically the same as a tangible idea of extension. He is now going to claim, based on his account of how we see distance, size, and situation, that it is possible to show there are no common sensibles, only objects proper to each sense. This broader thesis is the subject of the final section.

What Berkeley does not say, either in the final section or when he first announces his project, is why he intends to show there are no common but only proper sensibles. In the course of *The New Theory*, in fact, he never says why he thinks this is an important or interesting thesis to establish. In the *Theory of Vision Vindicated*, however, in two separate passages, he takes up the issue of common sensibles and says some things that indicate why he took it to be important to address this issue. In the first, *TVV* 15–16, Berkeley says:

> 15 It hath indeed been a prevailing opinion and undoubted principle among mathematicians and philosophers that there were certain ideas common to both senses: whence arose the distinction of primary and secondary qualities. But I think it hath been demonstrated that there is no such thing as a common object, as an idea, or kind of idea perceived by both sight and touch.[3]
>
> 16 In order to treat with due exactness on the nature of vision, it is necessary in the first place accurately to consider our own ideas: To distinguish where there is a difference: To call things by their right names: To define terms, and not confound our selves and others by their ambiguous use: The want or neglect whereof hath so often produced mistakes. Hence it is that men talk as if one idea was the efficient cause of another: Hence they mistake inferences of reason for perceptions of sense: Hence they confound the power residing in somewhat external with the proper object of sense, which is in truth no more than our own idea.

Berkeley holds the notion that there are common sensibles to be widespread and to be responsible for the prevailing distinction be-

3. It is perhaps worth pointing out that this paragraph makes clear that a common object is a kind of idea and not a kind of thing.

tween primary and secondary qualities. The suggestion is also present that he holds this notion to blame for some of the issues that have been important to the argument of the *New Theory*, such as that of confusing the "inferences of reason" with "perceptions of sense."

The second passage occurs at *TVV* 41:

> As to light, and its several modes or colours, all thinking men are agreed that they are ideas peculiar only to sight; neither common to the touch, nor of the same kind with any that are perceived by that sense. But herein lies the mistake, that, beside these, there are supposed other ideas common to both senses, being equally perceived by sight and touch, such as extension, size, figure, and motion. But that there are in reality no such common ideas, and that the objects of sight, marked by those words, are intirely different and heterogeneous from whatever is the object of feeling, marked by the same names, hath been proved in the *Theory*, and seems by you admitted. Though I cannot conceive how you should in reason admit this, and at the same time contend for the received theories, which are as much ruined as mine is established by this main part and pillar thereof.[4]

Here Berkeley tells us he regards the issue of the heterogeneity of the ideas of sight and touch to be what separates his theory from all others. All other theories are committed to the claim that we both see and touch extension, figure, and motion, whereas his theory is grounded on the fact that the extension, figure, and motion we see is of a different kind than the extension, figure, and motion we touch.

This description of the theories Berkeley regards as his rivals is somewhat puzzling. The commonly accepted theory against which Berkeley has been arguing is not, after all, a common-sense view that perception is a matter of straightforwardly apprehending a physical object unproblematically available to both sight and touch. Rather, what the received theories of perception Berkeley has been criticizing hold is that what is perceived by sight is a construct of the visual system that represents the physical object that is the cause of what we see. It is not at all clear what the geometrical theorists would

4. The *Theory of Vision Vindicated* was written in response to a letter in the *Daily Post-Boy* of September 9, 1732, attacking the thesis of the *New Theory of Vision*, which Berkeley had recently reprinted as an appendix to *Alciphron*. "You," therefore, refers to the author of this letter.

say about the role of touch with respect to the visual object. Descartes does, of course, make comparisons between what is learned by sight and by touch that suggest he thinks the same sort of calculations occur in both instances. But his point is that what we see, the information received in vision, like the information received from touch, need not resemble its physical cause. There is no suggestion we could use touch to supplement vision in apprehending a common object.[5] What distinguishes Berkeley's theory from his geometric rivals is that Berkeley's theory makes use of tangible ideas in order to supplement visual ideas in explaining how we see distance, size, and situation, whereas the geometric theorists make no mention of touch at all but explain our ability to see distance, size, and situation entirely as a process carried out by the visual system.

What it is, in fact, that makes Berkeley's theory different from its geometric rivals is that he claims that touch provides us with a different *sort* of information, which therefore makes it plausible to take touch to be available as a supplement to the visual system. Berkeley is then able to argue that distance, size, and situation are, in themselves, perceivable by touch and not by sight. We perceive distance, size, and situation by sight because visual signs, which are not in and of themselves ideas of distance, size, or situation, suggest to us tangible ideas that are habitually associated with them. The procedure of the geometric theory, on the other hand, assumes a single notion of extension, figure, and motion, which is what makes plausible the suggestion that the connection between what is immediately registered and what is mediately perceived is necessary and thus an appropriate subject of calculations. The calculations assigned to the visual system move indifferently between a visual construct, immediately registered information that may well be tangible such as the impact of light rays on the retina, and a mind-independent physical object. The geometric theory, moreover, will

5. It is not at all clear that this conclusion is one geometric theorists, and Malebranche in particular, would be likely to draw. Malebranche's argument has been that even in the case of extension, figure, and motion, the calculations of the visual system are not error-free, because they depend upon the peculiarities of the system, the way information is registered. There is no reason to suppose that Malebranche thinks that the tactile system is error-free, since he takes vision to be the most perfect of the senses, or that he would have any reason to believe that the errors of touch echo the errors of vision, so that we would end up with the same construct.

ignore touch as a supplement to the visual system so long as touch and sight are regarded as different ways of making sensible the same abstract or mind-independent properties of extension, figure, and motion. A geometric theorist could agree with Berkeley that we don't see and touch the same thing, but only in the sense that the sensible realization of extension in vision is different from the sensible realization of extension in touch. The geometric theorist would differ from Berkeley in holding that both sensible realizations represent and bear necessary connections to the same mind-independent or intelligible extension.[6] This difference does then underlie an important or crucial difference between Berkeley's picture and that of his rivals. Berkeley holds that the sensible ideas of sight and touch are of very different sorts and can bear only contingent relations with each other. This is what allows him to claim that what we perceive is not a representation of something to which it is necessarily connected, but instead, that what we perceive are the visual and tangible qualities themselves, which, when habitually found together, make up objects.

Extension-in-Abstract (NTV 122–126)

Berkeley, therefore, says that, before turning directly to the question of common sensibles, it will be "proper" to take up extension-in-abstract (*NTV* 122). What blinds others to the possibilities of tangible ideas for explaining space perception is a belief in common sensibles, a belief that the content of what is perceived by sight is the same as the content of what is perceived by touch. What in turn supports this belief, according to Berkeley, is not part of a theory of perception at all but rather what he calls a "secret supposition" that it is possible to form an idea of extension that has been abstracted from all sensible qualities, either visual or tangible. Berkeley's claim is that an unexamined assumption has become incorporated into a theory of perception. The theories rival to his are committed to the view that it is possible to form an intelligible idea of extension which is an

6. What I have in mind here is of course the rather more explicit formulation of Malebranche on this topic.

idea of what extension is like in and of itself in its own nature when it is not being perceived. Such an idea of mind-independent extension can be abstracted from all of the conditions of sensory perceiving, from whatever content of a perception is due to the nature of the particular sensory system involved. It is Berkeley's contention that the assumption there are common sensibles is based on the conviction we can have an idea of extension-in-abstract. What is assumed is that both sight and touch can provide sensory realizations of the same mind-independent extension about which we frame intelligible ideas and hence that ideas of sight and touch share a common content. It therefore becomes central to Berkeley's case to show that we don't, in fact, have an idea of extension-in-abstract—in order to refute an unspoken argument that it is the existence of extension-in-abstract that accounts for the presence of common sensibles.[7] He can then move on to remind his readers that previous arguments have already shown there are no common sensibles.

Berkeley's argument here against the possibility of framing an idea of extension-in-abstract is the first occurrence of an argument he repeated in several different versions.[8] This version, though

7. This unspoken argument in fact occurs in Malebranche's *Dialogues on Metaphysics*, a work there is no evidence Berkeley had read. In Dialogue V, Malebranche has Theodore say: "Let us suppose that you are looking at the color of your hand and at the same time feel pain in it. You would then see the color of the hand as extended and you would at the same time feel the pain as extended. Do you not agree with this? Aristes: Yes, Theodore. Moreover, if I touched it, I would also feel it as extended; and, if I plunged it into hot or cold water, I would feel the heat and cold as extended. Theodore: Note this then. Pain is not color, color is not heat, nor heat cold. Now, the extension of the color—or joined to the color—which you see when you look at your hand is the same as the extension of the pain, the extension of the heat, the extension of the cold which you are also able to sense. Hence, the extension belongs neither to the color, nor to the pain, nor to any of your other sensations. . . . It is then, the idea or archetype of bodies which affects us in different ways. That is to say, it is the intelligible substance of Reason which acts on our mind through all powerful efficacy and which affects and modifies the mind with color, taste, pain, by what there is in it [the intelligible substance of Reason] which represents bodies." *Dialogues on Metaphysics*, trans. Willis Doney (New York, 1980), p. 111.

8. Berkeley's most elaborate version of this argument occurs in the Introduction to the *Principles concerning Human Knowledge*. Other versions can be found in *3D* I, *A*, VII, 5–7 (1st and 2d eds.) (Wks., vol. 3, pp. 331–35), and *A Defence of Free Thinking in Mathematics*, Wks., vol. 4, pp. 45–48. I have discussed Berkeley's account of abstract ideas in "Berkeley's Anti-Abstractionism," in *Essays on the Philosophy of George Berkeley*, ed. Sosa. For some other recent accounts of Berkeley's anti-abstractionism, see Ken-

succinct, is in several respects the clearest. Berkeley describes the idea against which he is arguing as follows:

> We are therefore to understand by extension in abstract an idea of extension, for instance, a line or surface intirely stript of all other sensible qualities and circumstances that might determine it to any particular existence; it is neither black nor white, nor red, nor hath it any colour at all, or any tangible quality whatsoever, and consequently it is of no finite determinate magnitude: For that which bounds or distinguishes one extension from another is some quality or circumstance wherein they disagree. (*NTV* 122)

Such an idea as Berkeley describes does bear resemblances to, for example, the idea of intelligible extension as described by Malebranche. On Malebranche's account, any knowledge we have of the existence of bodies derives from a confused sensation together with a clear or pure idea of extension. He says: "We see or sense a given body when its idea, i.e., when some figure composed of intelligible and general extension, becomes sensible and particular through color or some other sensible perception by which its idea affects the soul and that the soul ascribes to it, for the soul almost always projects its sensation on an idea that strikes it in a lively fashion" (*ST*, Tenth Elucidation, p. 626) Anything in such an experience that is particular is due to sensation: "The sensation of color that the soul ascribes to figures makes them particular, because no modification of a particular being can be general" (*ST*, Tenth Elucidation, p. 625). Since whatever is particular and determined in a representation is due to sensation, such features are perceiver-dependent and cannot represent the body as it exists external to us. Knowledge of the external body will be contained only in abstract ideas, as extension-in-general or triangle-in-general, which present the range of possibilities of which the extension of triangles is capable. To the extent, then, that what we perceive can have external reference, it must

neth Winkler, "Berkeley on Abstract Ideas," *Archiv für Geschichte der Philosophie*, 65 (1985) 63–80, Martha Brandt Bolton, "Berkeley's Objection to Abstract Ideas and Unconceived Objects," in *Essays on the Philosophy of George Berkeley*, ed. Sosa, George Pappas, "Abstract Ideas and the '*esse* is *percipi*' Thesis," *Hermathena*, 139 (1985) 47–62, and Thomas M. Lennon, "Berkeley and the Ineffable," *Synthese*, 75 (1988) 231–50.

embody an idea of intelligible extension. Thus, on Malebranche's view, the mind-independent can be represented by general ideas whose content differs from perceiver-dependent sensations because they contain nothing determined or particular. It is this sort of view, which finds within an act of perceiving both a particular and determined sensation and an abstract intelligible idea, that Berkeley is attempting to refute. He is denying it is possible to distinguish the content of the abstract idea from the content of the particular sensation.

The grounds for Berkeley's denial are that such an idea can't be framed. He says:

> A line or surface which is neither black, nor white, nor blue, nor yellow, etc., nor long, nor short, nor rough, nor smooth, nor square, nor round, etc., is perfectly incomprehensible. This I am sure of as to my self: how far the faculties of other men may reach they best can tell. (*NTV* 123)

The approach that Berkeley is taking here has often been found confusing. He seems to be basing his rejection of the possibility of framing abstract ideas on his own inability to do so. But this in and of itself does not seem to be a good reason for claiming that Malebranche, for example, suffers from a similar disability. Berkeley's argument, however, has a more general application. He thinks he is unable to form such an abstract idea because the abstract idea would lack any content. His broader intuitions are that an idea which lacked any finite determinations is an idea that is contentless. If you remove from an idea of a line or a surface any way in which the line or surface is determined as a particular line or surface, any way in which some one line or surface can be distinguished from all other lines or surfaces, then you have also removed from the idea of the line or surface any way in which it can be extended. Determined lengths or colors are the ways in which lines or surfaces as we experience them take up space or are extended. You can't remove from the idea all determined lengths or particular colors and have any extension left at all. In the actual conditions in which we are aware of the qualities of extended bodies, we are aware of them as determined in the various ways perceivable by our sensory apparatus. Being colored is the way in which the things we experience take up space visually. What Berkeley is objecting to is a view that says that a conception of what mind-independent extension is like can be identified with the idea we are left with when we abstract from all the particular conditions under which we sense external bodies. When all sensory determinants like color and shape are annihilated, we are left not with an idea of mind-independent extension but with no idea at all.

Berkeley turns briefly at this point (*NTV* 124) to a subject he will return to at length later in this section. Many people, he says, think geometry studies abstract extension. It should be clear that such a belief underlies the attempt to explain space perception in terms of geometric reasoning. Such an attempt requires the assumption that the same property is being measured whether we are talking about the mind-independent extension of the cause of what we see or about extension as realized in sight or touch. Berkeley's notion is that what supports such an assumption is the conviction that geometry studies abstract extension that can be realized in a variety of ways. But, he says, geometry is the science that demonstrates how figures are to be measured. It is possible to measure only a finite, determined magnitude. An idea of an abstract figure, like that of a triangle in general, is one from which all finite determined magnitudes have been removed. Such an idea, therefore, would be of a figure that could not be possibly be measured and hence could not be the subject of geometry.[9]

Berkeley thinks this point, that an abstract idea of a figure would be one that couldn't be measured, is illustrated by Locke's discussion of the abstract idea of a triangle. Berkeley quotes a passage from Locke's *Essay* (4.7.9) in which Locke describes a general idea of a

9. Berkeley also mentions that abstract ideas are held to be what make universal propositions possible, but that it would be easy to show, "did I think it necessary to my present purpose," that a proposition can be universal without being abstract. Berkeley spends more time in later discussions of abstract ideas emphasizing that he is not opposed to the view that particular ideas can be made general, only to the notion of abstraction. (See the discussion, for example, in *A Defense of Free Thinking in Mathematics*.) What should be clear is that Berkeley does not think that the present discussion hangs on abstract *general* ideas. It is abstraction, separating the content of an idea from its context, and not generality that is bothering him.

triangle as one which "must be neither oblique nor rectangular, neither equilateral, equicrural, nor scalenum; but all and none of these at once. In effect, it is somewhat imperfect that cannot exist; an idea wherein some parts of several different and inconsistent ideas are put together." Berkeley's citing of Locke's triangle has had a somewhat unfortunate effect on later discussions of his theory of abstraction, since it has tended to give rise to the impression that Berkeley is interested in how we frame general ideas rather than, as I have been suggesting, concerned with the assumption that it is possible to distinguish within an act of perceiving some content that is abstractable from perception and some that is not. Thus Berkeley's point about Locke's triangle is often thought to attribute to Locke the view that an idea of a triangle that represents all triangles is one that contains the properties of all triangles, and Berkeley's reasons for rejecting Locke's view is then taken to be that such an idea cannot be represented or pictured in the mind's eye.

Berkeley's alleged argument against Locke has been treated with varying degrees of respect and is often said to misrepresent Locke's views, but I think it fails to capture what was actually bothering Berkeley about Locke's triangle. What Berkeley is picking up on is Locke's claim that the abstract idea of a triangle is of "something imperfect that cannot exist." The reason why the idea of the triangle is imperfect is because it omits all those features which distinguish one triangle from another, like a determined length to its lines or a determined size to its angles, and it cannot exist because a triangle can't exist without having lines of a determined length or angles of a determined size. Berkeley, in the passage in the *New Theory*, emphasizes that such an idea, on Locke's own grounds, ought to be recognized as inconceivable. He thinks once Locke has recognized that the idea of a triangle without any determined lengths or sizes is an idea that is inconsistent, an idea of something that cannot exist, then Locke ought to say, as Berkeley wants to say, that it is inconceivable, since Berkeley can quote Locke as committed to the view that what is inconsistent cannot be conceived. (Berkeley cites *Essay* 3.10.33.) It is clear why Berkeley thought the conjunction of these two passages from Locke's *Essay* ought to be enough to put an end to the possibilities of framing ideas that rely on abstraction. On the other hand,

Berkeley makes clear that the mistake against which he is arguing is not Locke's alone. He says:

> That a man who laid so great a stress on clear and determinate ideas should nevertheless talk at this rate seems very surprising. But the wonder will lessen if it be considered that the source whence this opinion flows is the prolific womb which has brought forth innumerable errors and difficulties in all parts of philosophy and in all the sciences. (*NTV* 125)

This more general mistake, I have been arguing, is that of assuming we can find an intelligible idea of extension embedded in an idea of objects whose properties are determined in sensory experience. It is this mistake on which the possibility of common sensibles can be supposed to hang. Common sensibles are thought to share the same content, but this common content is taken to be abstractable from all those particular conditions determined in perception.

No Common Sensibles (NTV 127–138)

Berkeley has gotten rid of what he regards as the reason why most people assume there are common sensibles, namely, that there is such a thing as abstract extension whose presence accounts for the existence of common sensibles. He is now going to argue there are no common sensibles. In *NTV* 127, he, in essence, throws down the gauntlet on this point: "*The extension, figures and motions perceived by sight are specifically distinct from the ideas of touch called by the same names, nor is there any such thing as one idea or kind of idea common to both senses.*" Berkeley's arguments in favor of this claim are extremely terse, and they have been regarded as unsatisfactory.[10] The reason for their terseness, however, is that Berkeley, as he says, regards his point about the nonexistence of common sensibles as having already been established in earlier sections. The function of his current arguments is therefore to summarize and to remind his readers of im-

10. See, for example, Pitcher, *Berkeley*, pp. 53–58.

portant points he can already count on. When the arguments are fleshed out appropriately, their nature and impact become a little clearer.

Berkeley's first argument in *NTV* 128 that there are no common sensibles relies on what he has already established in his various discussions of the Molyneux Man. To say there are common sensibles is to say that an idea of sight, a visual quality, or something we learn about by seeing is recognizably of the same sort as an idea of touch, something we learn about through feeling or touching. If an idea I am now having is of the same sort with previous ones, it cannot be entirely new to me. There must be something I can point to that will indicate why a current experience deserves the same name as some old experience. These remarks about when two ideas are of the same sort are quite vague, as George Pitcher complains. It is true, however, that it is not Berkeley who is claiming some ideas of sight and of touch are of the same sort. Therefore in this context it is not Berkeley who owes us an account of how to recognize when two ideas are of the same sort. All Berkeley has to do is to point out that anyone who is resting a great deal on this claim that there are common sensibles has to be able to point to *something*—whatever it might turn out to be—that can serve as the basis for such a claim. Berkeley's reason for pointing this out is because he thinks that his earlier arguments about the Molyneux Man have "clearly made out" that a Molyneux Man, accustomed to operating with tangible information alone, would find his new visual experiences to be entirely new to him and without any content shared with or similar to his old tangible experiences.

Berkeley's claim, therefore, is that, because of what has been already worked out with respect to the Molyneux Man, we already know there can be no common sensibles, because ideas of sight and touch are of different sorts. This means he thinks what he has established when he asks us to consider what the Molyneux Man knows about distance, size, and situation is enough to have already shown there are no common sensibles. It would be helpful here to go over what Berkeley thinks can be shown with an argument about the Molyneux Man and distance perception, to take one case. This will help clarify just what are the elements of the argument Berkeley thinks he can count on. When Berkeley asks us to think about what the experience of a Molyneux Man is like with respect to distance, he thinks we know, because is it widely agreed that anyone, including a Molyneux Man, can immediately perceive light and colors by sight. We also know that what we immediately perceive by sight bears no necessary connections with the concept of distance, with how far away the things we are seeing are. We know this because the arguments refuting the geometric theory have shown that the lights, colors, and muscular cues of the visual system are not registered as lines and angles and hence cannot serve as the preliminary data of a calculation about how far away things are. The ways these cues are registered, as lights, colors, and muscular cues, are conceptually unrelated to distance information. So since what the Molyneux Man sees are lights and colors, and since, moreover, he tells how far away things are by reaching out and touching them or by moving toward them, and since what he sees has no conceptual connection with distance, then what he sees has no conceptual connection with the kinesthetic and tangible means by which he immediately registers distance. It is an obvious and clear conclusion that the Molyneux Man couldn't tell just by looking how far away the things he is seeing actually are.

If, as it happened, the Molyneux Man could tell upon first being restored to sight how far away the lights and colors he is now able to see are from him, this would be only because there is some common connection between what he is now seeing and the way he customarily tells by touch how far away things are. But what thinking about the situation of a Molyneux Man allows us to see is that this is not the case, that the information derived by sight, the visual qualities that are learned about in seeing, is of a different nature and conceptually unconnected with the tangible qualities he already knows about. Thinking about the experience of a blind man restored to sight brings into focus that there is a tangibly based way of telling how far away things are which the blind have at their disposal and which is equally available to the sighted. Hence we know what sorts of qualities are available to the Molyneux Man through touch, and we know what sorts of additional qualities he will come to learn about by sight, and the nature of our knowledge is such that we must conclude that the visual qualities he experiences on being restored to sight will be entirely new to him and do not resemble the tangible qualities with which he is familiar. His visual experiences are new and are not of the same sort as his tangible experiences, and so there are no sensibles of the same sort and common to sight and touch.[11]

For this account of Berkeley's procedure to make sense, he has to be assumed to be conducting a thought experiment at this point. He has to take himself to be in a position to be able to assume we already have enough information to know ideas of sight are of a different sort from ideas of touch. This means he does not really think the question of the existence of common sensibles is up for empirical grabs. Much of the resistance to Berkeley's Molyneux Man arguments proceeds on the assumption that Berkeley is talking about a matter that can be settled empirically. Pitcher, for example, says that for the experience of a Molyneux Man to be a successful refutation of the claim Berkeley is trying to defeat, we have to assume the Molyneux Man, like the normally sighted, will have visual experiences that are clear, sharp, and well-defined. Pitcher doubts a man who has just been operated on for cataracts, which is to say, a man under those circumstances in which the congenitally blind typically can be given sight, would have experiences of this sort. But since the question of the Molyneux Man is to be settled by a *thought* experiment, not an operation, we can assume anything we like about his new visual experiences (and, indeed, can consider his sight restored in any way we like).

It is quite true that an actual Molyneux Man can't settle the question of common sensibles, if this is taken to be an empirical question, but he couldn't settle it either for or against Berkeley's claim.[12] He can't settle the matter in favor of Berkeley if he can't recognize by sight matters known to him tangibly because, as Pitcher points out, it is impossible to tell how much of an inability to recognize by sight what was previously known by touch is due to malfunctions in the visual system. He can't settle the matter against Berkeley

11. The arguments with respect to the size and situation judgments of the Molyneux Man proceed very similarly. We know what sort of size or situation information the Molyneux Man has by virtue of touch, and we know what kind of information he will acquire when he comes to see. Since we know there are no connections, either necessary or resembling, between these two sets of qualities, we know that the Molyneux Man's visual experience will be of something entirely new to him.

12. Armstrong also argues that empirical evidence is inconclusive on this matter. See *Berkeley's Theory of Vision*, pp. 62–65.

either, even if he could tell what something known to him by touch, like a cube (to revert to the original example), looked like when he first saw it. This is because a real Molyneux Man is not actually in the position of a newborn learning Berkeley's language of nature. He knows the sighted use the same words for things they learn about through sight and through touch. So the question *this* Molyneux Man is going to be asking himself is Which of my new visual experiences is more likely to be a representation of some tangible experience? Because Berkeley thinks that a visual square is a better sign for tangible squares than a visual circle, Berkeley himself might not be discomforted if an actual Molyneux Man got the answer to Molyneux's classic question about the globe and the cube right. But be that as it may, Berkeley himself can't take the truth of his claims to await research on real-life Molyneux Men.[13] He takes it instead to rest on arguments demonstrating the lack of necessary connections between the tangible means by which we experience distance and the visual cues that merely suggest distance to us. The function of the Molyneux Man is merely to help us pull apart and conceptualize something we already know about from our own experience: the differing content of the tangible means and the visual cues.

Berkeley's second argument, put starkly in *NTV* 129, hangs on the claim that the *only* immediate objects of sight are lights and colors. Since anyone will agree lights and colors are not the sorts of things perceivable by touch, anyone will also have to agree there is nothing perceivable by sight of the same sort as something perceivable by touch. Berkeley recognizes in *NTV* 130, however, that most people, indeed, "even amongst those who have thought and writ most accurately concerning our ideas,"[14] will not agree that the only qualities perceivable by sight are light and color because they also think that we can immediately see "space, figure, and motion."[15] Berkeley's

13. What Berkeley says in his appendix to the second edition of *NTV* about the William Jones case is that "if any curious person hath the opportunity of making proper interrogatories to him thereon, I should gladly see my notions either amended or confirmed by experience." In *TVV*, he says that the report of the Chesseldon case, from which he quotes, confirmed what he "had been led into . . . by reasoning" (*TVV* 71).

14. Berkeley quotes Locke, *Essay* 2.9.9.

15. This is Locke's list in the passage Berkeley quotes.

argument centers on figure or extension, since he says he has already shown that space or distance is not an immediate object of sight. His argument depends on the inconceivability of an idea of color abstracted from visible extension. He doubts "whether it be possible for [someone] to frame in his mind a distinct abstract idea of visible extension or figure exclusive of all colour: and on the other hand, whether he can conceive colour without visible extension" (*NTV* 130). Since Berkeley thinks this sort of abstraction is impossible, he also thinks it cannot be said that we perceive both color and extension: "In a strict sense, I see nothing but light and colours, with their several shades and variations" (*NTV* 130).

Pitcher is unhappy with this argument because he thinks the premise Berkeley is trying to establish, in order to be able to claim that visible extension is not tangible extension, is that color is identical with visible extension. He assumes Berkeley is trying to argue that whenever two qualities are inconceivable apart from each other, then the two qualities are the same as each other. Pitcher, I think quite correctly, finds this claim unconvincing; his example is that, if it were true, we would find ourselves saying that triangularity is the same as trilaterality. What is not clear, however, is that Berkeley either intends to make or needs to make a claim about identity. It is unlikely this is a claim that he would accept, since he is prepared to say it is possible to pay selective attention to one of two qualities that cannot be conceived apart, that it is possible to think about color without thinking about or paying attention to visible extension, and this surely would not be the case if he thought color was the same as visible extension. So it is not clear Berkeley intends to claim that color is identical with visible extension.

It is not clear either that this is a claim Berkeley needs to make in order to establish his argument. Pitcher has Berkeley arguing that since visible extension is the same as color, then visible extension is not the same as tangible extension. But we do not find Berkeley making this claim about visible and tangible extension. All he actually seems to be saying is that we don't have two separate ideas, of color and of extension, and so we can't claim that among the objects of sight we find both color and extension. (Instead, we see color patches.) In order to argue we don't perceive ideas of two separate qualities, color on the one hand and extension on the other, all

Berkeley needs to be able to claim is that they are mutually inextricable; that is, when we see a color patch, we have no way of determining which of what we see is what the color would be like in the absence of extension and which is extension in the absence of color. Their mutual inextricability allows Berkeley to argue that we will not be able to frame an idea of what color is like when it is not extended or what visible extension is like when it is not colored, the claim he is actually trying to establish. It is significant that the first part of this claim is one that was widely held by the geometric theorists Berkeley is criticizing and their contemporaries. It was often said that color is a mode of extension because color cannot be conceived without extension, but that extension is not a mode of color because extension can be conceived without color. Berkeley's point is that the latter feat is actually no more possible than the former, that the existence of visible extension depends as much on the existence of color as color depends upon extension. It is equally true to say that being visibly extended is a way of being colored as it is to say being colored is a way of being extended. Berkeley's argument therefore is, in the end, going to encourage us to substitute a picture of a set of mutually independent sensible qualities for a picture of a hierarchy of modes dependent on substances. It is this first picture which allows him to say that "by the mediation of light and colours other far different ideas are suggested to my mind: but so they are by hearing, which besides sounds which are peculiar to that sense, doth by their mediation suggest not only space, figure, and motion, but also all other ideas whatsoever that can be signified by words" (*NTV* 130).

Berkeley's third argument against the existence of common sensibles, in *NTV* 131, is that if there were an extension common to sight and touch, it would be possible to add together a visible line and a tangible line to reach a common sum. But since such addition is impossible, a visible line is not of the same sort as a tangible line. Armstrong's reaction to this argument is perhaps most natural. He asks But why can't I add these lines together? Berkeley's argument relies on a challenge to perform a mental feat, that of adding a visible to a tangible line. Only if you can actually do so is it correct to call the two lines of the same sort. Berkeley is so confident this cannot be done that he feels he can simply say: "I leave it to the reflexion and experience of every particular person to determine for himself" (*NTV* 131). Berkeley's argument seems unsatisfactory because he doesn't say why he is so sure the lines can't be added.

The actual example Armstrong gives, however, shows it is not so easy to demonstrate that the visible line and the tangible line can be added together. Armstrong asks Why can't I pace out part of a distance and then measure the rest by eye? It is true this piece of measuring seems perfectly possible, but what is questionable is whether it can be described as adding a tangible to a visible line. Berkeley can suppose it has already been thoroughly established that there is no line visible out from the eye but only lights and colors. When I estimate by eye how much distance is left after I have got done pacing, what I am actually doing is estimating, from the way things look, how much tangible distance is left. To add a tangible line to a visible line, I have to have paced or felt a distance between two tangible points and then add that distance to a line in the visual array, that is, to a line terminated by visible points, which can be seen to take up a definite number of points or minima in the visual field. The reason why these two lines cannot be added together has already been established in the section on situation perception. Adding the lines together requires the use of a measuring device whose units are neither visible nor tangible, and such a device, for the reasons just discussed in the course of the second argument, is inconceivable.[16]

In sections 132–136, Berkeley returns to a consideration of the Molyneux Man. He introduces this section by saying: "A farther confirmation of our tenet may be drawn from the solution of Mr Molyneux's problem, published by Mr Locke in his *Essay*" (*NTV* 132). It seems Berkeley is now turning to a fourth argument against the existence of common sensibles. The earlier argument of *NTV* 128 concerning the Molyneux Man had exploited Berkeley's own quite extensive conclusions about what could be learned from thinking about how a Molyneux Man would deal with distance, size, and

16. On the account I have been giving, it is not strictly true to say, as Pitcher does, that the only reason Berkeley has for claiming that a visible line can't be added to a tangible line is that visible lines are "in the mind." It is quite true that Berkeley holds the experience of visible lines to be mind-dependent, but his argument that visible and tangible quantities are incommensurable would continue to hold, even on his mature view that experiences of tangible lines are also mind-dependent.

situation. Now Berkeley is pointing out a conclusion that can be drawn from *Locke's* treatment of the problem for which the Molyneux Man was introduced: Molyneux's famous question Can a blind man upon being restored to sight recognize at first sight a globe and a cube he had been accustomed to distinguishing by touch?[17] Locke says he agrees with Molyneux "that the blind man at first sight would not be able with certainty to say which was the globe which the cube, whilst he only saw them" (*Essay* 2.9.8, quoted in *NTV* 132). Berkeley's point is that if Locke's and Molyneux's solution to the question whether the blind man can recognize the globe and the cube on first being made to see is correct, then this solution is inconsistent with the position, which he has also quoted Locke as holding, that distance, figure, and motion are common sensibles. For if the visual experience of a globe were of the same sort as tangible experiences of globes, then the ideas of each would have to have something in common that justifies their being called by the same name, and the Molyneux Man would be able to recognize the globe at first sight by means of this element shared by the visible and the tangible experiences of a globe.

Berkeley follows up this point with a reminder in *NTV* 135 of the way that the account of the nature of visual experience developed in the section of situation perception (especially *NTV* 106) bears on his approach to the Molyneux problem. What the Molyneux Man would have called a globe or a cube before his sight was restored to him would have been a family of tangible properties. What he perceives upon being made to see is not an organized visual representation, which might or might not be thought to bear connections or resemblances with the familiar collection of tangible properties. Instead, what the Molyneux Man experiences is an enormous number of different visual qualities, and he has not yet learned anything about which are likely to recur together, let alone which are likely to recur in the presence of his recognizable collection of tangible qualities. So the question Which of what you are now seeing is a globe and which a cube? would be unanswerable by him. His visual

17. I am disagreeing with Pitcher's treatment of Berkeley's discussion of the Molyneux Man. Pitcher runs together the arguments of *NTV* 128 and *NTV* 132–133 into a single argument, which he calls, contrary to Berkeley's numbering system, the second argument.

experiences aren't at present *of* anything, and he could no more solve this task than he could, for example, tell which of two odors would be likely to belong to which of two entirely unfamiliar plants.

In *NTV* 136 Berkeley addresses an objection that is often raised against his claims about Locke's and Molyneux's solution to the Molyneux problem.[18] Pitcher, for example, argues it ought to be possible to agree with Locke and Molyneux that the Molyneux Man would have trouble recognizing the globe and the cube at first sight without giving up the claim that visible squareness and tangible squareness are the same sort of quality. Pitcher says: "The fact that he [the Molyneux Man] knows how squareness *feels* should establish no antecedent expectation that he ought instantly to know how it *looks*" (p. 56). Berkeley argues that if the two experiences are thought to have something in common, then the fact that this something in common is being apprehended through different senses shouldn't prevent the Molyneux Man from recognizing this alleged common content to his different ideas of sight and touch. Berkeley says: "For though the manner wherein it affects the sight be different from that wherein it affected his touch, yet, there being beside this manner or circumstance, which is new and unknown, the angle or figure, which is old and known, he cannot choose but discern it" (*NTV* 136). It is helpful, in understanding why Berkeley's intuitions are so different from Pitcher's, to keep in mind that the issue being discussed by Berkeley is that of common *sensibles*. Berkeley is asking Do the ideas arrived at through different senses have anything in common? Is there any common content to our sensory ideas? He is not asking Do we perceive a common thing by different senses? Thus what Berkeley is trying to compare is how squareness feels with how squareness looks. His question is Do these different experiences have anything in common? So if Pitcher is prepared to agree that how squareness feels is no clue to how it looks, then, from Berkeley's point of view, he is agreeing that this is not a case of experiencing a common sensible.[19]

18. The fact that Berkeley addresses this objection is not often acknowledged.

19. Armstrong makes a similar response to this objection as made by Warnock, without, however, quoting the passage in which Berkeley himself gives the same answer to the objection.

Berkeley's final argument concerns the perception of motion. His procedure in showing there are no common sensibles has been to show that none of the ordinary candidates for common sensibles, such as appear on the list quoted from Locke's *Essay*, actually present content apprehended by both sight and touch. Although in his previous four arguments Berkeley has concentrated on the perception of extension, he turns briefly now to complete his project by taking up motion. His argument again depends upon points previously established, in this case upon points reached by a consideration of the Molyneux Man in the course of the discussion of situation perception. Berkeley claims there is no common content to a visual and a tangible experience of motion because a Molyneux Man, familiar with motion apprehended by touch, would not be able to recognize anything in his new visual experience as being in motion in this tangible sense. This is because the perception of motion presupposes directionality: something moving is moving up or down or right or left. But directionality is immediately perceived kinesthetically, through the experience of reaching out and touching. The Molyneux Man's new visual experiences do not appear to be at a distance, so that they can be reached for, and thus there is nothing in what he now immediately sees that has any connection with directionality. The shifting colors of the Molyneux Man's new visual experience contain no content that would be perceived by him as moving.

Berkeley's arguments (or argument summaries) that there are only objects proper to each sense have been designed to support a particular picture of what perception is like. Pitcher, at one point, accuses Berkeley of treating "the mind as a kind of spiritual see-er, or viewer, that 'looks at,' and 'sees,' the objects of awareness that come before it" (p. 56). But this picture is, in fact, much closer to the views Berkeley is trying to reject than it is to the view he holds. On Berkeley's account, each sense organ presents us with or informs us about an independent set of qualities. These qualities bear only contingent connections to qualities belonging to other sense modalities; they are not connected either through necessary conceptual relations or by relations of resemblance. When Berkeley says that what we see is proper only to seeing and contains no content in common with what we touch, we can conclude there is nothing in the content of what we see which determines that it represent something that can also be represented by touch. Our task, when presented with some visual experiences, will not be to work out necessary connections or resemblances with something else the visual experiences might be deemed to represent. Some qualities, as visual qualities, come to represent other qualities, as tangible qualities, only when they have been observed to recur together.

Berkeley is arguing against a view that takes perception to be a matter of constructing "objects" that can be related to other perceptual "objects" by a spiritual see-er. Berkeley's claim that there are no common sensibles, only objects proper to each sense, is a step in his attempt to propose a new view of the nature of perceptual representation. Berkeley is not therefore claiming an apple perceived by sight is a different kind of thing from an apple perceived by touch. What we perceive by sight are always just visual qualities, light and color. These visual qualities are "of" or represent an apple when they are such as to have recurred with tangible and other qualities. Since it is all of these qualities, as a collection, that are what we call an apple, then what sensible qualities represent is not something nonsensory but other sensible qualities. This view of perceptual representation is, according to Berkeley, best captured by thinking of what we are understanding in perception as if it were a kind of language rather than a kind of geometric demonstration.

11

The Language Model, the Geometric Model, and Common Sense

In the *Theory of Vision Vindicated*, Berkeley tells us he is going to be giving his argument in an inverse order from that which he used in the *New Theory*. He will now proceed "analytically," by adopting the conclusion of the *New Theory* as a premise and deducing the rest of his argument from this premise. The principle that he proposes to adopt as his starting premise is: "That *Vision is the Language of the Author of Nature*" (*TVV* 38). The first theorem he draws from this principle is that connections between signs in the language of vision are arbitrary: "In fact, there is no more likeness to exhibit, or necessity to infer, things tangible from the modifications of light, than there is in language to collect the meaning from the sound. But, such as the connexion is of the various tones and articulations of voice with their several meanings, the same is it between the various modes of light and their respective correlates; or, in other words, between the ideas of sight and touch" (*TVV* 40). The characteristic connection between ideas in a language is arbitrary; there is no necessary connection or resemblance between a sign and what it signifies. So if vision is a language, then the connection between visual signs and what they signify, ideas of touch, must be arbitrary. Ideas of sight will be of an entirely different sort from ideas of touch.

The Language Analogy (NTV 139–148)

In the *New Theory*, Berkeley has spent a great deal of his time establishing this very point, that ideas of sight and touch are heterogeneous and bear no necessary connections or resemblances to each other. He now proceeds to establish what in the *Theory of Vision Vindicated* he had called the conclusion of the earlier work. He is going to argue that since the connections between the ideas of sight and touch have been shown to be arbitrary, then vision is best regarded as a kind of language. The way he goes about developing this demonstration is, however, a little less than fully straightforward. He says what he is going to do is to answer a series of possible objections to the thesis that ideas of sight and touch are of different sorts. But what his various replies to the different possible objections he puts forth allow him to do is to develop a plausible case for the claim that the arbitrary connections between the ideas of sight and touch indicate that vision functions like a language.

Berkeley considers three objections that might be put forward against his claim that there are no common objects of sight and touch. The first two objections allow him to use his answers to develop an analogy between the ideas of sight and touch and the words of a language and what the words stand for. The third objection allows him to deal with what looks like a disanalogy between vision and language. The first objection he raises (*NTV* 139) asks why it is a universal custom to call visible extension and visible shapes by the same name as tangible extension and tangible shape, if visible and tangible shapes are not the same sort of thing. Berkeley's answer (*NTV* 140) is that just as we call visible and tangible squares by the same name, so we call the written mark 'square' by the same name as the tangible square it stands for—but no one would want to say that a written mark on a page is the same sort of thing as a tangible object. In the case of language, we call the mark by the same name as the thing it stands for, because the mark, as a visual shape, has no intrinsic interest whatsoever. Our interest in it is solely as a means by which we can refer to tangible squares. Since our interest in visible shapes like visible squares is likewise primarily as a sign for what things feel like, it is natural to call the visible signs, which are uninteresting in themselves, by the same name as the tangible objects they signify. Thus this objection allows Berkeley to make the point that vision, like any other language, has a significatory function. What vision is for is to stand for nonvisible objects.

The second objection Berkeley takes up (*NTV* 141) says that a tangible square is nevertheless much more like a visible square than it is like a visible circle. So it seems reasonable to say that the tangible square represents the visible square because it resembles, and hence is of the same sort as, the visible square. In his answer Berkeley, rather surprisingly, concedes the major point of the objection. He agrees that the visible square is, as he says, "fitter" than a visible circle to represent a tangible square, but not for reasons that go against his claim that the relationship between a visible square and a tangible square is arbitrary. The reason why the tangible square is best represented by the visible square and not the visible circle is because the visible square has parts that correspond to the parts of the tangible square. But this is by no means to say that any of the parts of the tangible square resembles the corresponding part of the visible square. A visible and a tangible square can each have four sides without its being the case that a visible line or a visible angle be like a tangible line or tangible angle. Again we can observe that this is how a written language functions. The written characters have only an arbitrary connection with the sounds they stand for, but the language must be designed so that each sound has a corresponding written designation. "It is indeed arbitrary that, in general, letters of any language represent sounds at all: but when that is once agreed, it is not arbitrary what combination of letters shall represent this or that particular sound" (*NTV* 143). It is inappropriate to expect a symbol system to resemble what it represents. But just because resemblance is not an appropriate criterion by means of which symbol systems can be characterized does not give a reason for saying there are no criteria at all by means of which symbol systems succeed or fail, nor does it mean that one symbol system is as good as another for symbolizing some domain. With respect to vision, then, it is inappropriate to claim that vision fails as a representational system if visual signs don't resemble the things they represent. Vision can still be taken to be a well-designed and successful means of representation.

The final objection Berkeley takes up (*NTV* 144) concerns what looks like a disanalogy between vision and language. In the case of language, we show no inclination to suppose the signs are of the same sort with what they signify, but this is a mistake to which we are prone in the case of vision. We do confuse ideas that are actually tangible with visible ideas. Berkeley thinks he can account for the disanalogy, because in the case of languages, such as a written language or music notation, we can remember learning the language and so have no inclination to confuse the written word 'square' with the tangible ideas it stands for or to confuse a visual shape like a note on a page with the sound to which it refers. But visible signs have been constantly connected, from our earliest days with the tangible ideas they have come to signify. "We cannot open our eyes," Berkeley says, "but the ideas of distance, bodies, and tangible figures are suggested by them. So swift and sudden and unperceived is the transition from visible to tangible ideas that we can scarce forbear thinking them equally the immediate object of vision" (*NTV* 145). It is for this reason we need thought experiments like that of the Molyneux Man in order to capture an experience not ordinarily accessible to us, that of not knowing what a visible sign stands for.

Berkeley's answer focuses on a way in which the language of vision differs from other languages, namely, the language of vision is always with us. Thus his answer can seem to be a good explanation why we don't suppose, in the case of a language like music notation, that there is any natural resemblance or connection or identity between the visual shape that occurs in the notation and the sound, like middle C, for which the notation stands. Music notation functions very differently from the experience of seeing and feeling a cube. There the relevant visual experiences are universally and constantly conjoined to the same tangible experiences, whereas we remember learning music notations, which vary from culture to culture and whose signs appear in conjunction with their referents only when put to use by skilled readers. All this is at best, however, an explanation for why we take visible cubes to resemble tangible cubes, even though we don't make similar resemblance claims for other signs and their significants. Berkeley has made further claims about the relationship between visible and tangible ideas which, to many, have seemed quite unlike any other language. For the mistake I make when I see and feel a cube is that of supposing the two experiences are alike. I do not make the mistake of supposing I am seeing something I am actually feeling. Berkeley's account of the perception of distance, size, and situation appears to be resting on just that claim, however, that I am really feeling distance, size, or situation but I think I am seeing it. His account of the difference between vision and other languages in no way seems to explain what indeed seems incredible, that the deliverances of one sense modality could come to appear to be registered by a different sense modality entirely.

This apparent difficulty for Berkeley ignores, however, the extent to which he can take the language analogy seriously with respect to vision. The tangible ideas which get attached to visible signs are analogous to those ideas that get attached to written or spoken words when they are understood, to what, perhaps somewhat incautiously, might be thought of as the meaning of the word or at least those ideas which make a word meaningful to a speaker of the language. Now it is true that understanding the ideas that make a word meaningful requires a familiarity with particular experiences that the word designates. But an episode of understanding is not the same as undergoing the requisite experience itself. Indeed, it need not even be the same as reproducing the experience in imagination. Berkeley frequently makes the point that we often use words as signs without calling to mind the ideas for which they stand.[1] In general, however, I could not understand the word 'square' unless I *could* call to mind, that is, imagine, the tangible experiences for which the word stands and which constitute my idea of a square. But even when I am hearing the word with understanding, the only immediate sensory experience I am having is auditory, even though I ignore the strictly auditory properties of the experience when I hear the word as meaningful. In the same way, if I mediately see how far away something is, then, as a result of what I see, I understand or have received some distance information. The only sensory experience I am having is visual, but this experience is rendered meaningful by other ideas. These ideas, if I am to understand them, have to be realized in tangible or kinesthetic terms, since this is the only way

1. Berkeley makes this point for the first time in the Introduction to the *Principles*, where he writes: "And a little attention will discover, that it is not necessary (even in the strictest reasonings) significant names which stand for ideas should, every time they are used, excite in the understanding the ideas they are made to stand for: in reading and discoursing, names being for the most part used as letters are in *algebra*, in which though a particular quantity be marked by each letter, yet to proceed right it is not requisite that in every step each letter suggest to your thoughts, that particular quantity it was appointed to stand for" (*PHK*, Intro. 19).

distance is immediately experienced and represented by me. But understanding the visual ideas as standing for ideas of distance does not require undergoing the experience of distance. This is what makes the confusion of modalities possible. Distance information has been for so long and so constantly available to the eye that the viewer ceases to realize that the reason why visual experiences are understandable is because of other nonvisual experiences.

In *NTV* 147, Berkeley draws the conclusion his replies to his various objections have hinted at. It is this paragraph which contains what the *Theory of Vision Vindicated* had identified as the conclusion of the *New Theory*. It runs as follows:

> Upon the whole, I think we may fairly conclude that the proper objects of vision constitute an universal language of the Author of nature, whereby we are instructed how to regulate our actions in order to attain those things that are necessary to the preservation and well-being of our bodies, as also to avoid whatever may be hurtful and destructive of them. It is by their information that we are principally guided in all the transactions and concerns of life. And the manner wherein they signify and mark unto us the objects which are at a distance is the same with that of languages and signs of human appointment, which do not suggest the things signified by any likeness or identity of nature, but only by an habitual connexion that experience has made us to observe between them.

This paragraph sums up the points Berkeley made in his various replies. If vision works like a language, then visual signs do not represent through resemblance or necessary connections but instead by means of contingent connections established in experience. Thus, for Berkeley, the language analogy functions as an alternative to the geometric analogy on which the theories of Descartes and Malebranche are based. We cease to think, for example, of seeing where something is as a working out of a geometric problem and instead take it to be a matter of understanding visual signs. We judge the successes and failures of vision as a language, not, as the geometric theorists do when they talk about the errors of vision, by how well what we see resembles what it represents but instead by how well it performs the functions of a language. The function of a language is to signify, the function of vision as a language is to signify that by means of which we can regulate activity. Vision works well to the extent that it permits us to guide our actions to achieve our various ends.

Berkeley concludes this section with a paragraph in praise of vision. In many ways, the sentiments expressed in the paragraph would be familiar to readers of his day. It was not at all uncommon to single out vision for special remark as an example of God's skill and ingenuity as a designer. Berkeley puts the point this way:

> The wonderful art and contrivance wherewith it is adjusted to those ends and purposes for which it was apparently designed, the vast extent, number, and variety of objects that are at once with so much ease and quickness and pleasure suggested by it: All these afford subject for much and pleasing speculation, and may, if any thing, give us some glimmering, analogous praenotion of things which are placed beyond the certain discovery and comprehension of our present state. (*NTV* 148)

This attitude toward vision, which among other things requires a willingness to employ a teleological approach toward the workings of nature, contrasts sharply with the assessment of vision provided by Malebranche. Malebranche develops an account of vision that enables him to show how unsuccessful vision is as a source of information about the natural world. Berkeley's theory enables him to stress how well vision works. It does exactly what it is supposed to do: by revealing visible properties that are habitually connected with tangible properties, we are able to use what we see as a guide to our actions as we interact with and operate on the tangible properties of the natural world. Thus, in defining the function of vision as he does, Berkeley is also reasserting the primacy of the sensible world as the object of our knowledge of nature.

The Objects of Geometry (NTV 149–159)

In the final section of the *New Theory*, Berkeley considers what is the proper subject matter of geometry.[2] He says he is not in general going to draw any corollaries from his theory, but he cannot resist answering this question about geometry. The importance of settling

2. See also the discussion by Armstrong, *Berkeley's Theory of Vision*, pp. 58–59.

questions about the nature of geometry for the success of Berkeley's overall theory is obvious. It was the success of geometry as the basis for the science of optics that led to the application of geometry as a model for vision. It is this geometrical model Berkeley has proposed to replace with a language model. What has been undeniably successful has been the discovery that a great deal of the mechanics of vision can be explained simply by considering the geometrical relations that obtain between the object to be seen, the light rays, and the various organs of vision, as the lens and the retina. The success of this approach to optics had led to the development of a psychological theory of vision in which the process of seeing is modeled on a geometric calculation. Seeing is thought of as a two-stage process in which the geometric properties of what we immediately see serve as input for a geometrical construction of a three-dimensional visual representation of the geometric properties of the object to be seen, as embodied in certain specifically sensible properties like color. Berkeley says he is now going to ask the question Is the object of geometry something visible or something tangible? Sorting this matter out is going to provide the final nail in the coffin of the geometric theory of vision. The support derived for the geometric theory from the geometric science of optics presupposes that geometry studies a common or abstract object, no matter what sense modality is involved. This "secret supposition" has already been dismissed. Berkeley is now going to show that he is entitled to conclude that geometry can appropriately be applied to the tangible objects studied by the physical science of optics but not to the objects of vision. Geometry, therefore, can play no role in the psychology of vision, although it has a proper role in the study of optics.

Answering his question is actually going to require nothing more of Berkeley than reiterating the conclusions he has reached previously, but he goes over his former reasoning because the answer Berkeley favors is not the one he believes most people find intuitively obvious. He thinks, in general, people are inclined to say what geometry studies is visual in nature, because they take the subject of geometrical reasoning to be the visible figures seen on a page. But visible figures have no fixed size to be measured and hence cannot be the subject of geometry. Instead, we should recognize that the visual figures seen on a page function like words to suggest tangible shapes and sizes. Tangible shapes and sizes, on the other hand, can be measured, and hence geometry studies the tangible and not the visible properties of things.

This means Berkeley will have no desire to reject the use of geometry in optics, to the extent that what is being measured can be understood as tangible properties, as he showed could be done in his discussion of the inverted retinal image. In the *Theory of Vision Vindicated*, Berkeley claims as a virtue for his theory that it is able to claim "we may nevertheless compute in optics by lines and angles" (*TVV* 32) without being committed to the false theory that we see by lines and angles. Berkeley is able to preserve computation in optics while denying it a place in vision by separating the subject matter of geometry, the consideration of "particles as moving in certain lines, rays of light as refracted or reflected, or crossing, or including angles" (*TVV* 43), from the subject matter of the study of vision, which explains how perceivers see what they do. The content studied in the one case is limited to tangible properties, whereas the latter is, of course, concerned exclusively with how we register visible properties.

Berkeley completes his account by showing that in no sense can what we see, the proper objects of vision, be the subject matter of geometry. This argument constitutes his final demonstration that there is no room, within the psychological science of vision, for the space-measuring concepts of geometry. He asks us to consider the case of what Berkeley calls an "unembodied spirit."[3] The unembodied spirit is the tactile counterpart of the blind; such a person is assumed to be able to see but not touch. A real-life realization of such a person might be someone born with a paralysis of all motor functions and lacking a kinesthetic sense but with a working visual system. Such a person would have normal vision, in the sense of immediately seeing what every other sighted person can see, but could not feel and could not sense motion. Berkeley's case, however, does not rest upon any empirical evidence derived from or tested by

3. Such a person has been referred to in the literature, as by A. D. Ritchie, *George Berkeley: A Reappraisal*, ed. G. E. Davie (Manchester, 1967), as a "bodiless cyclops," 'cyclops' presumably because the bulginess of the visual field that is the result of binocular vision has been confused with metric depth perception, how far away the front of an object is from the back.

the existence of such a person; it is again in the nature of a thought experiment. The person Berkeley asks us to envisage, however, is equipped in the manner of the kind of see-er studied by a geometrical theory of vision, inasmuch as a geometrical theory assumes the perception of space can be accomplished solely by calculations of the visual system without reference to any other modalities, such as touch. The geometric theory treats the see-er as if it were an unembodied spirit.

Berkeley's claim is that the unembodied spirit would be unable to frame any of the geometrical concepts for measuring space and spatially organized objects. It is relatively straightforward to argue (as Berkeley does in *NTV* 154) that a spirit who couldn't sense movement couldn't do solid geometry. Without the experience of moving from place to place, the unembodied spirit would lack a concept of distance. The spirit would have no way of registering metric distance away from itself, how far away one place is from the place at which it is located. Without the concept of distance that derives from being able to move through space, it wouldn't be able to take anything it saw to be at a location out from or independent of itself, that is, it would lack the concept of outness; nor would it be able to take some parts of what it saw to be farther away than other parts, that is, it would lack the concept of depth.

The implication of Berkeley's claim is that being able to move and touch or, more precisely, being able to register movement and touching through kinesthetic and tangible sensations is what is required in order to be able to understand and apply the concepts of solid geometry. Being able to measure the spatial properties of solid objects requires no more than kinesthetic and tangible information. If we ask whether the blind can do solid geometry, the answer is yes, because the blind can register movement out into space and can register tangible sensations. The kind of question Berkeley is asking, when he asks about the possibility of an unembodied spirit's doing solid geometry, is a question that is answered by talking about what kinds of information are available to a spirit that has or lacks one or another sort of faculty or sense modality. It is not a question that requires any sort of presupposition about the nature of mind-independent space, because it is a question about the psychological endowment of perceivers. There is no reason, therefore, to expect

Berkeley's approach to geometry to change when, in the *Principles* and other works, he is more explicit about the mind-dependent nature of tangible ideas. Berkeley's discussion of what is needed in order to do geometry assumes that what are required are perceiver-dependent capacities.

Berkeley's perhaps more difficult claim is that the unembodied spirit couldn't do plane geometry either, because the spirit would also be unable to apprehend the movements that constitute measuring or defining objects on a plane surface (*NTV* 155). Since the spirit could not manipulate a rule or compass, it would lack the concept of right angle and circle, and since it could not carry out operations like imposing one angle on another, it would have no means for recognizing the equality of angles and hence would lack this concept. For Berkeley, geometrical demonstration consists in the manipulation of objects in space, placing one on top of another or making a circle with a compass. The success of a geometrical demonstration consists, therefore, in the successful completion of mechanical operations such as interposition. Even though we typically take an interest in the visual figures that are produced, what allows us to draw conclusions about the measurement of objects is the actual movements that produce fixed or determinate objects for which visual objects can stand.

The figure on a page either has a fixed size that can be measured by tangible operations (which is to say, has a tangible size) or has no determined size at all, because it can be viewed from any distance. As a visible object, it lacks fixed or measurable size. When we suppose the object of geometry is the visible figure, what we see when we look at what has been drawn on a page, we are making the same kind of mistake Berkeley exposed in the discussion of situation perception, that of taking the visual field to be like a picture in which the various color patches have the kind of spatial organization that is derived from being laid out on a single plane surface. This model, Berkeley has argued, attributes to vision a greater organizational capacity than is in fact the case. As Berkeley says, in *NTV* 157, what we see by eye is not uniform but multiform. What we see are just different visual qualities, lights and colors, which are not spatially structured at all with respect to each other, either on one plane, that is, all at the same distance, or at different distances. It is true that lights and colors are perceived in color patches, which are distinguishable one from another and which can therefore be judged to be greater or smaller one than another. But since the comparative sizes of color patches alter from moment to moment, very little is gained by comparing them. The extensions of the colors we immediately see are not experienced as structured on a single plane, one beside the other. In order for us to see what is before us as part of a single flat surface, the colors have to be reliably coordinated with the tangible experience of a flat surface, all of whose parts are equidistant. As Berkeley says: "We see plains in the same way that we see solids, both being equally suggested by the immediate objects of sight, which accordingly are themselves denominated plains and solids" (*NTV* 158). Thus the unembodied spirit is actually suffering a much greater disability than the congenitally blind. The blind have the same tactile and kinesthetic abilities as the sighted and can use them to get an idea of the organization of their spatial surroundings and to frame ideas of collections of tangible qualities making up kinds of tangible objects. But the unembodied spirit would not only be completely lacking in spatial concepts; without the fixed tangible meanings by means of which visible ideas are organized into collections, it would also, in a very real sense, be unable to learn to see or, more precisely, it would be unable to learn how to understand what it can see. It could, of course, apprehend visible qualities, but these visible qualities, lacking a means of organization as a representation, wouldn't be taken to be qualities of anything. They would, for the unembodied spirit, fail to represent.

If, as Berkeley has argued, the proper subject matter of geometry does not include what we see, then the geometric theory of vision is trying to solve a false or nonexistent problem. The geometric theory conceives the problem how perceivers see distance, size, and situation to be a problem about how it is that perceivers, on the basis of what they immediately see, are able to come to see a three-dimensional representation having spatial properties like extension, size, and shape. It will be recalled that Berkeley shares a theory of immediate perception with geometric theorists. They are agreed that we understand what is immediately perceived by sight by understanding what the visual system is equipped to register, and they are agreed that the full range of what we learn about by seeing is much greater than what we immediately perceive. Berkeley has shown, however, that on the basis of what we immediately perceive it is impossible to do the kind of geometry required in order to construct a three-dimensional visual representation. Therefore the geometric theory has misconceived the problem of how we see distance, size, and situation. This cannot be a problem about how we perceive a three-dimensional visual representation, for, in point of fact, a three-dimensional representation cannot be constructed on the basis of visual qualities and is therefore not the sort of thing that can be seen. Berkeley is not questioning the fact that we do indeed take what we see to be located at a particular situation in space, at a given distance, and of a fixed size. What he is questioning is whether this is a fact exclusively about the visual system.

Berkeley's preferred theory is that we take what we see to have distance, size, and situation because we have learned what these tangible qualities look like. We have learned to recognize the characteristic visual appearances associated with the tangible qualities of distance, size, and situation. Thus the language model, unlike the geometric model, doesn't require each sense modality to do anything other than what it is equipped to do. What we see is never anything more than the proper objects of vision and what we touch is never anything more than the proper objects of touch. The virtue of Berkeley's theory is that he can show, simply by assuming the proper functioning of the various sensory modalities, how it is that we can perceive distance, size, and situation.

Heterogeneity and Common Sense

The presence of the heterogeneity thesis in Berkeley's theory of vision has been singled out as an example of Berkeley's notorious willingness to fly in the face of common sense. Pitcher, for instance, concludes his discussion of the *New Theory of Vision* with a summary of the various assertions he takes Berkeley to be committed to which contradict some of our deeply cherished beliefs.[4] Chief among them are that we never see and feel the same object; that what we see does

4. Pitcher, *Berkeley*, pp. 59–61.

not resemble and does not represent what we feel; and that there is not a single object like a moon or a tree but instead a visible moon, a tangible moon, and indeed a moon for every sense modality that can be affected by it. Pitcher's discussion of Berkeley's assaults on common sense involves, to one degree or another, what seem to me to be distortions of Berkeley's intentions with respect to the heterogeneity thesis. This is because Pitcher takes the heterogeneity thesis itself to be what grounds Berkeley's rejection of a common-sense conviction that sense perception is a matter of unproblematically seeing and feeling physical objects. But although there is a sense in which Berkeley does reject such a common-sense view of sense perception, his rejection is based not on the heterogeneity thesis but instead on notions that he shares with the theory he has set up as his rival, the geometric theory. Whatever break Berkeley is making with common sense rests on the theory of immediate perception Berkeley shares with many other thinkers of his day. The heterogeneity thesis ought more properly to be seen as a defense erected by Berkeley against the noncommonsensical claims of the theories to which he is opposed.[5]

Pitcher's first example of a cherished belief Berkeley requires us to give up is that "all of us are convinced that when we touch an elephant while at the same time looking at it, we feel and see the very same object" (p. 59). But, as I have argued previously,[6] when Berkeley says we never see and feel the same object, he is talking not about those collections of ideas to which we give the same name, as, elephant, but instead about sensory objects, contents of ideas. He is claiming the way we immediately see the elephant (the content of our visual ideas) is not the same as the way we feel the elephant (the content of our tangible ideas). This claim is already much more commonsensical. For most people are not firmly convinced the way the elephant looks, as, gray, is exactly the same as the way the elephant feels, as, rough. It is perfectly true Berkeley's account of this fact does result in a certain kind of challenge to common sense. This is because he thinks we can identify what we strictly or imme-

5. For another view of Berkeley's relation to common sense, see Pappas, "Berkeley, Perception and Common Sense."

6. See chapter 5.

diately see in terms of what can be registered by our sensory apparatus. We will find we actually see less than we think we see and what we see is unavoidably perceiver-dependent, dependent upon the perspective and perceiving apparatus of the perceiver. This narrowing of what we actually see has the result of challenging such common-sense beliefs as that we can see how far away something is from us or that we can see its size. But this theory, which recognizes the subjective nature of what we sense, is the expression of much of the idealism, the theory of ideas, characteristic of the seventeenth century and is by no means unique to Berkeley. Although accepting this theory, moreover, does require us to think more carefully about what we think we see, it is not clear that denying it puts us back squarely in line with common sense. To deny that in seeing there is some way in which we see which the perceiver apprehends would amount to a denial that we use our minds in seeing, and to deny that the way in which a perceiver sees reflects that perceiver's viewing apparatus amounts to denying we use our eyes in seeing. It is by no means clear that denying these theses amounts to robust common sense. On the contrary, even apart from the suggestion we can understand vision independent of any investigation of the nature of the visual system or visual processes, any way of expressing a claim that seeing is neither mental nor visual would seem to amount not to common sense but to nonsense.[7]

Because he accepts a theory of immediate perception, Berkeley is not going to want to suppose, with common sense, that we see just what's out there; that since things are at a distance, we just see that distance. Berkeley's own theory, however, is not directly critical of this common-sense conviction, since he does not take this to be a part of what is agreed by all theorists. Instead, the contribution of his own theory that is important to him, and is supported by the heterogeneity thesis, is the claim Pitcher says Berkeley "assaults us further with," that "the things we see do not even resemble, and are not representations of the things we feel" (p. 59). Now it is true that we do typically think, for example, that the size we see resembles the size we touch and represents the size of the object. But what, accord-

7. I am making the same kind of point for which Yolton has argued forcefully in, among other places, *Perceptual Acquaintance*.

ing to Berkeley, will force us to give up this belief is the theory of immediate perception, which shows us that the size we see can vary whereas the size we feel is fixed. Indeed, it is facts of this sort, recognized both by Berkeley and by his rival geometric theory, that result in the view that space perception is a two-stage process. Both theories hold that because what we immediately see does not resemble the distance, size, and situation perceptions we arrive at, what we see requires supplementation in order for us to be able to perceive spatial properties. Malebranche develops an account of the two-stage process in which space perception is a matter of calculations based on necessarily imperfect, because perceiver-dependent, sensory data. He claims that what we see does not resemble and hence imperfectly represents the object that is its cause and for which it stands. When Berkeley, on the other hand, asserts the things we see do not resemble the things we feel, he is trying to defend a common-sense belief that our senses accurately inform us of our surroundings. Because what we feel does not resemble what we see, tangible information can be understood as available to supplement visual information. And so long as we understand the representation relation to be contingent rather than necessary, to be based on habitual and not necessary connections, then visible ideas provide perfectly adequate representations of, or suggestions for, tangible ideas, which, according to Berkeley, are to be seen as their meanings. It is by giving up the resemblance claim that we can continue to accept the common-sense belief that what we see provides accurate cues to what we touch, so that we can, through the mediation of visual ideas, perceive tangible qualities like distance, size, and situation.

Finally, when Pitcher says Berkeley is committed to the view that there is no one single object, like a tree, but rather a visible tree and a tangible tree, he is again misusing Berkeley's term 'object.' What Berkeley is actually committed to is the view that the way we apprehend a tree visually is not the same as the way we apprehend the tree tangibly, and hence that what we identify as the visual qualities of the tree, the qualities apprehensible by sight, are not the same as its tangible qualities, the qualities apprehended by touch. It is the rather more conventionally representational account provided by a geometric theory that has as a consequence there is a visual tree, an organized visual construct that represents, but is not the same as, the tree. Berkeley's position is that the tree's visible qualities represent a tree only when they have been observed to coexist with those tangible and other qualities which collectively are what an idea of a tree stands for. Since it is Berkeley's position as well that being a tree amounts to no more than being a collection of coexisting sensory qualities, then Berkeley can claim that our senses are adequate to inform us of the existence and nature of trees. It is a well-known feature of Berkeley's later theory, as laid out in the *Principles*, that he can claim that, according to him, snow is white and apples are red, whereas this same common-sense claim must be rejected by materialists. Berkeley's project in the *New Theory* allows him to make similar sorts of claims about space perception. The space he is showing us we learn about by seeing is the very same space in which we move about and reach out and touch things, in the sense that what we see coexists with and is coordinated with the space we touch. The distance through which we move is how far away the object is from us, and the location to which we reach to touch the object is the object's spatial situation. Objects are not, as they are in the geometric theory, placed in imperceptible locations at imperceptible distances inaccurately represented by the constructs of vision, the most accurate of our inaccurate senses.

In general, Berkeley's heterogeneity thesis becomes far less of an affront to common sense when it is not read as being about different kinds of objects and is instead taken to be about the nature of our sensory systems. Although no one would want to maintain, I should expect, that a complete understanding of the nature of sensory processing can be derived from our various common-sense convictions on this topic, the fact remains that we hold some fairly deeply cherished beliefs about sensing and sensations. Among these beliefs are, on the one hand, that there is something inescapably subjective about sensing, that it is a way of apprehending that happens to a perceiver, and that the way perceivers perceive reflects their own perspective. We also believe that by using our senses we gain an accurate and usable knowledge of our environment. It is this set of common-sense beliefs that Berkeley can argue he has preserved against a rival theory that gives up the latter claim for the sake of the former. In his eyes, it is the heterogeneity thesis that permits him to do so, and hence it is the heterogeneity thesis that preserves common sense.

PART THREE

Some Implications

12

The New Theory *and Immaterialism*

Berkeley conceived his general approach to the theory of vision, as illustrated by his various solutions to the problems of distance, size, and situation perception, to rest on the view that "the proper objects of vision constitute an universal language of the Author of nature, whereby we are instructed how to regulate our actions in order to attain those things that are necessary to the preservation and well-being of our bodies, as also to avoid whatever may be hurtful and destructive of them" (*NTV* 147). It is Berkeley's contention that we will understand how it is vision works successfully as it does when we understand the representative function of visual ideas on an analogy with language. Visual ideas represent because they call to mind other ideas with which they are arbitrarily but customarily connected, in the way that written or spoken sounds call to mind the meanings that are arbitrarily but customarily associated with them. As is the case with language, successful seeing is something that must be learned. Perceivers have learned the meanings of their visual experiences.

Berkeley's theory of the nature of visual representation was intended to replace and indeed did replace an account in which what we see was thought to be a construct of the visual system, which represented a mind-independent object by virtue of necessary connections holding between the construct and the external object. These constructs were held to be not learned but the deliverances of a visual system that recapitulated in geometric principles its own causal history. An important difference between Berkeley's theory and this geometric theory is that according to Berkeley vision successfully represents, whereas, under the geometric approach, the visual construct only imperfectly represented the object for which it stood. Berkeley replaced a psychology of vision that supported skeptical conclusions about the senses with one in which vision worked.

In order to achieve this result, that vision works, Berkeley had to reconceptualize what it is like to gain knowledge of the natural world, through vision or by other means. In Berkeley's eyes, vision works when we are able to know what to expect on the basis of what we see. We are equipped, by our senses, to acquire knowledge of or to be aware of those qualities proper to each sense. Through sight, we are aware of visible qualities, through touch of tangible qualities, through smell of olfactory qualities, and so forth. These various sets of qualities exhibit no relations, either necessary or resembling, with the qualities proper to other sense modalities. Sensible qualities are heterogeneous. Our task, in dealing with the disparate information provided by the different senses, is to work out what experiences we can expect on the basis of the experiences we are now having. Since these experiences are conceptually unconnected, the way we do this is to look for regularities we can count on. In the cases under discussion in the *New Theory*, we learn through experience what visual qualities are reliably associated with what tangible qualities. We learn what tangible qualities look like. On this account, therefore, what our sensory experiences represent are other sensory experiences. They cannot be connected with and do not represent some mind-independent object on which our ideas are held to depend. We should not therefore take the function of our various senses to be that of enabling us to build or construct a representation of a mind-independent object, and we will not have any reason to be troubled by the fact that a construct based on mind-dependent sensations necessarily represents only imperfectly a mind-independent object.

This general account of what it is like to have knowledge of the natural world reappears in Berkeley's *Principles of Human Knowledge* and his *Three Dialogues between Hylas and Philonous.* In both of these later works, Berkeley draws the same positive lesson, that, according to his view of things and according to his conception of natural knowledge, knowledge of the natural world is entirely within our grasp. For example, in *PHK* 31, Berkeley writes:

> That food nourishes, sleep refreshes, and fire warms us: that to sow in the seed-time is the way to reap in the harvest, and, in general, that to obtain such or such ends, such or such means are conducive, all this we know, not by discovering any necessary connexion between our ideas, but only by the observation of the settled laws of Nature, without which we should be all in uncertainty and confusion, and a grown man no more know how to manage himself in the affairs of life, than an infant just born.

Berkeley conceives the *Principles*, like the *New Theory*, to have a positive moral. In the *Principles*, his aim is to combat skepticism (and atheism) by demonstrating that it is possible through our senses to reach a knowledge of the natural world. Just as the *New Theory* argued against the claim that vision is not suited to show us the nature of things, by rejecting a picture in which the role of vision is to provide a representation of mind-independent spatial properties, so in the *Principles*, Berkeley attacks the notion of mind-independent material objects or material substance directly and concludes that our senses are suited to inform us of the "settled laws of Nature" which constitute our knowledge of the natural world.

In a long passage toward the end of *Three Dialogues*, Berkeley again draws the conclusion that knowledge of nature is a matter of knowing the regularities that exist among ideas, relying on examples that are reminiscent of arguments found in the *New Theory*.

> Strictly speaking, Hylas, we do not see the same object that we feel; neither is the same object perceived by the microscope, which was by the naked eye. But in case every variation was thought sufficient to constitute a new kind or individual, the endless number or confusion of names would render language impracticable. Therefore to avoid this as well as other inconveniencies which are obvious upon a little thought, men combine together several ideas, apprehended by divers senses, or by the same sense at different times, or in different circumstances, but observed however to have some connexion in Nature, either with respect to co-existence or succession; all which they refer to one name, and consider as one thing. Hence it follows that when I

> examine by my other senses a thing I have seen, it is not in order to understand better the same object which I had perceived by sight, the object of one sense not being perceived by the other senses. And when I look through a microscope, it is not that I may perceive more clearly what I perceived already with my bare eyes, the object perceived by the glass being quite different from the former. But in both cases my aim is only to know what ideas are connected together; and the more a man knows of the connexion of ideas, the more he is said to know of the nature of things. What therefore if our ideas are variable; what if our senses are not in all circumstances affected with the same appearances? It will not thence follow, they are not to be trusted, or that they are inconsistent either with themselves or any thing else, except it be with your preconceived notion of (I know not what) one single, unchanged unperceivable, real nature, marked by each name: which prejudice seems to have taken its rise from not rightly understanding the common language of men speaking of several distinct ideas, as united into one thing by the mind. (*3D* III, pp. 245–46)

The particular conclusions of the *New Theory* appear to constitute a special case of the more general conclusions Berkeley takes himself to be establishing in the *Principles* and *Three Dialogues*. Berkeley's claim to have shown the way to refute skepticism in the *Principles* and *Three Dialogues* can be argued to have received support from his demonstration in the *New Theory* of how it is possible to have visual knowledge of spatial properties. The argument of the *New Theory* constitutes a case history, showing how to apply the general methods for which Berkeley argues in the later *Principles* and *Three Dialogues*. To the extent, moreover, that the *New Theory* is regarded as providing support for the *Principles*, its arguments can also be understood as guides to understanding the more comprehensive and also less readily acceptable arguments of the *Principles.* If the *New Theory* and the *Principles* are read as dedicated toward the same overall project, then the arguments of the *New Theory*, by means of which Berkeley brought about a revolution in the study of vision, can provide a useful tool for interpreting those claims of the *Principles* widely held to be incredible.

To conceive the argument from the *New Theory* in this way, however, as providing support for the argument of the *Principles*, it must be the case that the two works are mutually consistent. But, it is widely held, they are not consistent. The *New Theory*, Berkeley tells us, commits what he calls in the *Principles* the "vulgar error" of taking tangible objects to have a mind-independent existence. Therefore, it is often said, the *New Theory* is materialist in a way that is not the case for the *Principles* and *Three Dialogues*, or the *New Theory* subscribes only to an immaterialism of the visuals.[1] Although Berkeley's solutions to the problem of how we perceive by sight distance, size, and situation require that the content of what we see is mind-dependent, they also permit the comfortable assumption that the spatial properties suggested by our visual ideas are positions and extents of absolute, mind-independent space. In addition, so long as arguments in the *New Theory* claim to be consistent with and indeed incorporate the conclusions of geometric optics, they rely on the existence and causal powers of materialist entities, like light rays, which are not perceivable. This has seemed to mean that the *New Theory* relies on entities that are unacceptable in the starker ontology of the *Principles.* So, if it is indeed the case that the *New Theory* relies on an ontology different from and more acceptable than that of the *Principles*, the success of the *New Theory* as a theory of vision cannot be a reason for accepting or even understanding the full-blown immaterialism of the *Principles.*

This claim, however, that the ontology of the *New Theory* renders it inconsistent with the *Principles*, is contrary to Berkeley's own assessment of the relation between the *New Theory* and the *Principles*. The passage in the *Principles*, *PHK* 44, where Berkeley admits to the vulgar error, supposes the argument from the *New Theory* can, without loss, be expressed in terms of ideas and gives examples of what such a rewritten version would look like. What Berkeley says is this:

> The ideas of sight and touch make two species, entirely distinct and heterogeneous. The former are marks and prognostics of the latter. That the proper objects of sight neither exist without the mind, nor are the images of external things, was shewn even in that treatise. Though throughout the same, the contrary be supposed true of tangible ob-

1. See, for example, Pitcher, *Berkeley*, Warnock, *Berkeley*, and A. A. Luce, "Editor's Introduction" to *An Essay towards a New Theory of Vision*, in *The Works of George Berkeley, Bishop of Cloyne*, vol. 1 (London, 1948), pp. 143–57. See also Genevieve Brykman, *Berkeley, Philosophie et apologétique* (Paris, 1984).

> jects: not that to suppose that vulgar error, was necessary for establishing the notion therein laid down; but because it was beside my purpose to examine and refute it in a discourse concerning *vision.* So that in strict truth the ideas of sight, when we apprehend by them distance and things placed at a distance, do not suggest or mark out to us things actually existing at a distance, but only admonish us what ideas of touch will be imprinted in our minds at such and such distances of time, and in consequence of such and such actions. It is, I say, evident from what has been said in the foregoing parts of this treatise, and in Sect. 147, and elsewhere of the essay concerning vision, that visible ideas are the language whereby the governing spirit, on whom we depend, informs us what tangible ideas he is about to imprint upon us, in case we excite this or that motion in our own bodies. But for a fuller information in this point, I refer to the essay it self.[2]

Berkeley's claim is that there is nothing in the argument of the *New Theory* that requires the proper objects of touch "exist without the mind" or "are the images of external things." Berkeley apparently does not suppose his demonstration of the way in which we see spatial properties in any way trades on a supposition about mind-independent space, nor does he apparently take his references to the behavior of unobservable entities such as light rays to be inconsistent with the ontology of the *Principles*. Many of his readers, however, have taken a contrary view on these matters. The notion that the *New Theory* is inconsistent with the *Principles* has rested on the assumption that a version of the *New Theory* argument, rewritten in terms of ideas, does not provide a satisfactory solution to the problems of space perception and indeed, many have supposed, amounts to a denial of the existence of space or, at least, of distance. Before the *New Theory* can be regarded as support for the *Principles*, and before its arguments can be used as a guide to understanding the *Principles*, it is necessary to straighten out in what sense, if any, the *New Theory* is materialist and to what extent it is at odds with the immaterialism of the *Principles*.

2. A similar, rewritten version of the argument about distance occurs in *Three Dialogues*, where Berkeley says: "From the ideas you actually perceive by sight, you have by experience learned to collect what other ideas you will (according to the standing order of Nature) be affected with, after such a certain succession of time and motion" (*3D* I, p. 202).

Inconsistencies between the New Theory *and Later Works*

The upshot of the *New Theory* is that vision is a language standing for what, in the *New Theory*, Berkeley calls tangible objects, which he sometimes, as he admits, speaks of as if they existed without the mind. It is this same notion, that vision is a language, that I have argued Berkeley intended to carry over as the conclusion of the *Principles* and *Three Dialogues.* But in the transition from the *New Theory* to the *Principles*, Berkeley now wants to say that what visual ideas suggest to us are other ideas about how and in what circumstances we will be affected. In this transition, Berkeley has lost an asymmetry between vision, the content of which he identified as the signs, the wordlike part of the language, and touch, whose content supplied what was signified, the meaninglike part of the language. In the *New Theory*, even though the visual signs were so many ways in which the perceiver was affected, what the signs stood for could be understood to be mind-independent bits of the material world. Once Berkeley wants to claim an equivalent status for objects of sight and touch, once he asserts in both cases we are talking about ways in which the perceiver is affected, it becomes questionable whether he can continue to claim it is vision that supplies the signs of the language we learn and it is touch that supplies what is signified. Would it not instead be the case that touch is as much a sign for what we will see as vision is a sign for what we will touch? Is it not the case that, without the materialist underpinnings of the *New Theory*, the application of the language analogy to the ontology of the *Principles* is not immediately obvious or straightforward?

G. J. Warnock is an example of someone who thinks the notion of a 'divine visual language' cannot be easily transferred from the *New Theory* to the *Principles*.[3] Warnock argues as follows:

3. Warnock, *Berkeley*, p. 43. A. A. Luce, in his "Editor's Introduction" to the *New Theory*, also supposes that Berkeley must give up the asymmetrical sign-signifier relation in the *Principles*. Luce writes: "In this respect the second book has superceded the first, and was meant to do so. If it be true that colours, the immediate objects of sight, are signs of coming or possible objects of touch, it is no less true that roughnesses, etc. the immediate objects of touch, are signs of coming or possible objects of sight" (p. 150). Luce does not see this, however, as affecting the claims about a divine visual language.

> All through the *Essay* Berkeley takes for granted a sharp distinction between the sense of touch and the other senses. The real distance, size, and 'situation' of anything is determined solely, he says, by the sense of touch. Furthermore, the 'proper objects' of sight and hearing, taste and smell, are said to be 'not without the mind'—not (he here evidently means) actually at a distance from the observer. But what we touch, he asserts, really *is* at a distance; it really has quite definite size and shape; it is this that is really in 'circumambient space'. This, then, is why God's language is (predominantly) visual. What we see is a sign of what we could touch, that is to say, of what is really there; just as a man's name painted on a door indicates that the man himself is to be found behind it. As the man himself is not a sign, so also what we touch is not a sign. It is the real thing; and God's visual language merely serves to point it out to us.

As Warnock sees it, what makes the objects of the sense of touch uniquely suited to be what is represented by visual and other ideas is that these objects really are sitting at specific and mind-independent distances at specific locations in absolute space where they take up a specific extent, or really have a mind-independent size and shape. Once these claims are abandoned, then there is no reason to suppose that touch is uniquely suited, or indeed suited at all, to inform us of these spatial properties and hence no reason for claiming it is the objects of touch that are immediately represented by visual ideas.

Berkeley himself, however, never argues in the way Warnock suggests. He does not claim that what makes touch uniquely suited to represent spatial properties is that what we touch is real in a sense not true for what we see. His argument is rather that the content of what we see, lights and colors, is lacking in usable spatial information but that this is not true for the content of what we apprehend through touch. Reaching out and touching is a way of experiencing the distance and situation of things, and for this reason distance and situation are among those things whose nature is an immediate object of touch. Similarly, we learn about size and shape tangibly but not visually, because the tangible experiences of size and shape are stable and responsive to measurement, whereas there are no stable visual experiences of a shape or a size of some object. Finally, the reason why vision is a sign for touch and not the other way round is a pragmatic one: we find the stable properties revealed by touch to be more interesting than the fleeting visual properties and more important in our attempts to preserve our lives in a pleasurable rather than a painful condition. Therefore, none of the reasons Berkeley gives in the *New Theory* for taking vision to be a sign for what is revealed by touch depends upon a special ontological status for what we touch. Those reasons are instead based entirely on the content of the tangible as opposed to the visual experience and hence are completely translatable into, and indeed are already couched in terms of, talk of ways in which the perceiver is affected.

A second claim important to the *New Theory* which has been held to be inconsistent with the *Principles* is that the objects of sight and of touch are heterogeneous, of entirely different kinds. A. A. Luce, for example, argues that according to the ontology of the *Principles*, "visible and tangible are both sense data, as we should say, or ideas, in Berkeley's technique; they belong to the same genus, and so they are not, strictly, heterogeneous. They are both alike in the mind. Once the sensible thing is defined as 'a collection of ideas', the original doctrine of the heterogeneity loses its force and point; *visibilia* and *tangibilia* must be in the thing on an equal footing, and no metaphysical distinction in kind between them can be sustained."[4] In both the *New Theory* and the *Theory of Vision Vindicated*, Berkeley supports the claim that vision is a kind of language by arguing only on this supposition can we understand how it is we are able, through vision, to acquire knowledge about the entirely unrelated objects of touch. If Berkeley can no longer claim objects of sight and touch are heterogeneous, then he has lost a key element in his argument for the language analogy.

Luce's claim, however, that the heterogeneity of sight and touch is inapplicable to the ontology of the *Principles*, depends upon the assumption that when Berkeley says the objects of sight and touch are of entirely different kinds, what he is thinking about is that visual objects exist in the mind whereas tangible objects have an existence without the mind. If this were the form Berkeley's argument takes,

4. Luce, "Editor's Introduction," p. 151. Pitcher argues somewhat similarly: see *Berkeley*, p. 152. It is worth pointing out that in *PHK* 44 Berkeley says explicitly that the *ideas* of sight and touch are heterogeneous.

then there would, indeed, be severe disparities between the position of the *Principles* and that of the *New Theory*. But when Berkeley argues the ideas of sight and touch are of different kinds, he is basing this claim on the idea that there is no common content to what we experience by sight and by touch. The qualities we experience visually are entirely other than the qualities we experience tangibly, so that there are no means, except that of habitual experience, of working out connections between the two. The issue that concerns Berkeley to which the claim of heterogeneity is relevant is the nature of sensory representation. He is arguing against a theory which holds that visual ideas represent the objects on which they depend through resembling or necessary connections. Berkeley is trying to show instead that the visual qualities we perceive by sight represent by suggesting the entirely different sorts of tangible or other qualities with which they are habitually associated. Thus even in the *New Theory*, when Berkeley argues visual ideas represent the tangible, he is trying to build a picture of a sensible thing as a collection of unrelated sensible qualities.

The major difference between the *New Theory* and the *Principles* is one of emphasis. In the *New Theory*, Berkeley is trying to show that his theory of visual representation constitutes a better way of solving problems of visual space perception, and so he rests his claim on a heterogeneity or unlikeness principle in order to call attention to the suitability of tangible experiences for an account of space perception. In the *Principles*, on the other hand, he is interested in attacking directly the claim that sensory ideas represent mind-independent material objects, and so he makes use instead of a likeness principle, the claim an idea can be like nothing but another idea. Berkeley's gloss on the likeness principle when he first introduces it in the *Principles*, "a colour or a figure can be like nothing but another colour or figure" (*PHK* 7), suggests the basic notion behind the two principles is very similar[5] and certainly perfectly consistent. At any rate, since the heterogeneity principle, like the likeness principle, is about the content of experience, it too is expressible entirely in

5. If an idea can be like only another idea of the same sense modality, it will lack any common content with ideas from different modalities, and it will never be the case that two ideas from different sense modalities will be like something nonideational. See Turbayne, "The Influence of Berkeley's Science on His Metaphysics."

terms of ideas and cannot be seen as constituting a serious inconsistency between the *New Theory* and the *Principles*.

Perhaps the most critical inconsistency sometimes claimed to exist between the *New Theory* and the *Principles* concerns the importance of the distance argument. It is said the *Principles* ontology will no longer permit Berkeley to claim he has shown how it is we perceive distance by sight, since he is no longer in a position to claim there is anything at a distance the presence of which could be suggested by sight. The problems Berkeley successfully solved in the *New Theory* no longer exist in the framework of the *Principles*. This claim is nevertheless somewhat surprising in light of the fact that Berkeley continued to use versions of the distance argument, not only in *Principles* and *Three Dialogues* but in *Alciphron* as well,[6] and continued to cite the *New Theory* as showing we don't perceive absolute space or distance by sight. The claim doesn't seem to cohere very well with Berkeley's own perception of what he is doing.

That Berkeley can no longer claim to have shown how distance is perceived by sight is based on the assumption that if visual ideas, as is claimed in the *Principles*, merely put us in touch with tangible ideas, then, since tangible ideas, like visual ideas, exist only in the mind, they, like visual ideas, cannot exist at a distance. So it must be the case that, according to the *Principles*, there is no such thing as distance or space, nor could we actually move through space to encounter objects. This view has been expressed by I. C. Tipton as follows:

> It is true that when we read *NTV* in particular we are encouraged to think of persons as embodied agents who can reach out and touch things, and walk to and feel things, but ultimately Berkeley is going to hold, as he tells us in Pr. [*PHK*] 44, that this picture distorts the real truth. We have to shift our attention now to his beliefs about sensible objects. And when we do this we see that (strictly) there can be no question of our walking from A to B and feeling an object which was there to be felt, for the truth is supposed to be that there *is* nothing at a distance and that visual ideas are really only signs of what tangible ideas God is prepared 'to imprint on us.' In the last analysis I cannot learn about things by bumping into them, lifting them, or trying to

6. *PHK* 43; *3D* I, p. 202; *A* IV, 8 (Wks., III, pp. 150–51).

push them around, and this not just because a bodiless cyclops *cannot* manipulate objects, but because there are no manipulable objects. (p. 314)

Thus according to Tipton's view of the *Principles*, since there is no world of objects with spatial properties, as perceivers we can do no more than observe what is passing through our minds. Berkeley can no longer claim to have shown how we perceive distance by sight but only what makes it appear to us that we are so perceiving.

This view, however, that in the *Principles* Berkeley can no longer suppose that we can perceive distance, is based on a misunderstanding of the distance argument, which, again, mistakes the nature of the claim that visual ideas exist only in the mind. The reason why visual ideas are in themselves insufficient for perceiving distance and other spatial properties is not because they exist in some peculiar, nonspatial location, the mind. The reason why visual ideas are not immediately of distance is because the content of the ideas we are equipped to apprehend by sight is insufficient without supplementation to provide distance and other spatial information. It is also the case that these ways in which perceivers see, the proper objects of vision, are dependent upon their perceiving equipment and their perspective. It is in this sense that visual ideas do not exist without the mind; they are perceiver-dependent. Beings with our sort of visual apparatus register lights and colors but not distance. Thus when we tell by seeing how far away an object is, the distance information must actually be an idea from the imagination. An idea from the imagination is not, of course, an imaginary idea, merely one that properly belongs to some sense other than the one being exercised.[7] The content of tangible and kinesthetic experiences pro-

7. See, for example, *TVV* 9: "Besides things properly and immediately perceived by any sense, there may be also other things suggested to the mind by means of those proper and immediate objects. Which things so suggested are not objects of that sense, being in truth only objects of the imagination, and originally belonging to some other sense or faculty." It is sometimes said that Berkeley's position in *NTV* is that the distance that is perceived by sight is non-sensory (Brook, *Berkeley's Philosophy of Science*), or is always perceived mediately, even by touch (A. David Kline, "Berkeley, Pitcher, and Distance Perception," *International Studies in Philosophy*, 12 [1980] 1–8). *TVV* 9 makes clear this is not the case. Ideas that are perceived mediately by one sense are the proper objects, perceived immediately, by some other sense. (Although this is

vides an immediate understanding of distance and other spatial information. Moving through space, for example, *is* apprehending the distance between oneself and an object encountered. This is the way distance is experienced by perceivers capable of kinesthetic and tangible sensations. Thus even in the *New Theory*, the distances we perceive by touch are perceiver-relative, dependent upon our kinesthetic and tangible means of apprehending. Distance cannot be apprehended except through tangible experiences.

There is no reason, therefore, to suppose that Berkeley intends to change this part of his story within the framework of the *Principles*. Distance will continue to be immediately apprehended by touch. We will continue to have the kinesthetic experience of forging ahead uninterruptedly, hence through space, until we experience through touch a solid body. And to say distance or solidity can't be apprehended except through touch is not a denial of distance or solidity. What we experience kinesthetically is not the appearance of moving from A to B or the appearance of solidity but movement and solidity, just as what we see is not the appearance of color but color. It is far from being the case, as Tipton maintains, that in the *Principles* there are no movements and no manipulable objects. Doing things with one's hands, in the *Principles* as in the *New Theory*, is an important means of learning about objects. In the *Principles*, what Berkeley denies is the existence of nonmanipulable objects and of absolute space. He denies there are objects in a space which, being mind-independent and therefore imperceivable, remain beyond our grasp.

The claim to which Berkeley is committed is that distance and other spatial properties can be immediately experienced only by touch. This does not amount to a claim that distance and other spatial properties do not exist, any more than the claim that colors can be experienced only visually amounts to the claim that there are no colors. The only reason one might have for drawing such a conclusion would be if one supposed existence is always mind-independent, and hence that mind-dependent existence is always a

---

not what Berkeley says, it would seem plausible that the supplementary ideas of the imagination might also be ideas from the same sense, but which are not at the present moment being perceived.)

fictive substitute. This is not a supposition Berkeley is prepared to make. On the contrary, this is the claim Berkeley in the *Principles* and *Three Dialogues* has set out to refute. To say the items of our experience, like colors or distance, exist only as ideas is not to say, as ideas, they are representations of the nonexistent. This is the sort of picture that did indeed trouble Descartes and Malebranche. Since it is always possible, they reasoned, that we can be having an idea of the moon in the absence of a mind-independent object, the moon, the idea itself might always be representing the nonexistent. In Berkeley's eyes, we rid ourselves of these skeptical fears when we realize that sensible ideas are not the means by which we represent something mind-independent but merely mind-dependent ways of experiencing what exists.

In the *Principles*, just as in the *New Theory*, tangible objects, the objects of our sense of touch and kinesthetic sense, do exist at a distance, albeit an existence that is a perceiver-dependent. What they do not do, therefore, is exist without the mind. The contents of our tangible ideas are ways of feeling, just as the contents of our visual ideas are ways of seeing, and in neither case is it possible to abstract from what is a way of perceiving some content that continues to exist unperceived. This conclusion, it is true, is not drawn in the *New Theory*. Instead, Berkeley speaks as though what we perceive by touch could continue to exist unperceived. Nevertheless, none of the arguments of the *New Theory* is inconsistent with the mind-dependence of the objects of the sense of touch. None of the *New Theory* arguments designed to show that touch is a suitable supplement to vision in informing us about space trades on a claim that tangible objects are real in a way that is not true of visual objects or that they are out there in some mind-independent fashion. All such arguments are about the content of tangible information, and this content is perceiver-dependent. Even within the framework of the *New Theory*, distance is something immediately apprehended through touch and hence, as a way of perceiving, is perceiver-dependent, an idea. But this ideality of distance, its dependence on a perceiver, does not make it merely apparent, nor, certainly, does it render it nonexistent. At most, it is a reason for rejecting claims about mind-independent distance or about pure or mind-independent space. This is the conclusion that Berkeley is prepared to draw explicitly in the *Principles*. The presence of Berkeley's "vulgar error" in the *New Theory* does not constitute a serious inconsistency between the *New Theory* and the *Principles*. The perceiver-dependence of tangible objects is a conclusion available to be drawn within the context of the *New Theory*, even though Berkeley explicitly draws it only in the *Principles*.[8]

Corpuscularian Science and the New Theory

Another way in which the arguments of the *New Theory* have seemed to be at variance with the conclusions of the *Principles* has to do with the extent to which Berkeley is willing, in the *New Theory*, to adopt the results of materialist or corpuscularian science. R. J. Brook, for example, contends that the "*Essay* appears to abound, then, in materialist imagery" (p. 43). Brook cites passages from the *New Theory*, such as *NTV* 35, where Berkeley is prepared to talk about the geometric properties of light rays, and *NTV* 68, where Berkeley makes use of a description of the physical properties of light.[9] That is, although the *New Theory* proposes a new philosophy (or psychology) of vision, Berkeley claims as a virtue of his account that it can continue to explain the success of the existing optics or physics of vision. The *New Theory* does not propose an overthrow of the existing physics. But the *Principles*, with its attack on material

8. About this matter Turbayne says, "Accordingly, as far as the *Essay* is concerned matter still exists. But its character is remade. No longer is it beyond the reach of experience; no longer do we see pictures of it or the effects of it; no longer is it a subject or substance that supports attributes." "Editor's Commentary," in George Berkeley, *Works on Vision*, p. xliv. Armstrong also holds that accepting immaterialism does not require any changes in the *New Theory* with respect to the objects of touch but concludes that Berkeley was wrong in *PHK* 43 to hold that the two works were related.

9. *NTV* 68 reads: "Now in order to explain the reason of the moon's appearing greater than ordinary in the horizon, it must be observed that the particles which compose our atmosphere intercept the rays of light proceeding from any object to the eye; and by how much the greater is the portion of the atmosphere interjacent between the object and the eye, by so much the more are the rays intercepted; and by consequence the appearance of the object rendered more faint, every object appearing more vigorous or more faint, in proportion as it sendeth more or fewer rays into the eye."

substance, is generally held to do just that. The physics accepted in the *New Theory*, moreover, is to a large extent a physics whose ontology includes particles too small to be observed. But the *Principles* rests on the all-important tenet that 'to be is to be perceived', and this tenet has been thought to rule out the existence of bodies too small to be perceived.[10] So another way in which the *New Theory* appears to be materialistic lies in its willingness to countenance things like unobservable particles.

It is important to notice, however, that the objects whose existence is admitted in the *New Theory* are *not* materialistic in the sense rejected in the *Principles*. In the *New Theory*, the various entities that to Brook reflect a 'materialist imagery' are all specifically identified as tangible; they are apprehensible through touch. Berkeley spells this out unequivocally in the *Theory of Vision Vindicated*: "The retina, crystalline, pupil, rays, crossing refracted and reunited in distinct images, correspondent and similar to the outward objects, are things altogether of a tangible nature" (*TVV* 49). In the *New Theory*, Berkeley is not prepared to accept the existence of material objects in the sense against which he argues in the *Principles*. His theory does not require objects that, being independent of the senses, are as much intangible as they are invisible. On the contrary, his concern in the *New Theory* is to show that all talk of imperceptible causes of visual objects must be replaced by talk of things that are tangible. It is worth pointing out that, in *Theory of Vision Vindicated*, Berkeley is explicit that these tangible objects need not be such as to be actually touched by us. He says: "And here it may not be amiss to observe that figures and motions which cannot be actually felt by us, but only imagined, may nevertheless be esteemed tangible ideas, forasmuch as they are of the same kind with the objects of touch, and as the imagination drew them from that sense" (*TVV* 51). Berkeley's distinction is not between things that have been observed and things that have not but between those properties which are of the sort that can be experienced and those, such as are held to characterize mind-

10. For discussions of the issues surrounding this matter, see Daniel Garber, "Locke, Berkeley and Corpuscular Skepticism," in *Berkeley: Critical and Interpretive Essays*, ed. C. M. Turbayne (Minneapolis, 1982), pp. 174–93, and Margaret D. Wilson, "Berkeley and the Essences of the Corpuscularians," in *Essays on Berkeley*, ed. John Foster and Howard Robinson (Oxford, 1985), pp. 131–47.

independent objects, that are imperceptible-in-principle. This way of understanding Berkeley's distinction suggests that the dictum of the *Principles*, 'to be is to be perceived', is similarly intended not to distinguish between objects that are observed and objects that are too small or too remote to be observed but instead to distinguish between objects characterized by experientiable properties and objects whose properties are inexperientiable-in-principle.[11] In this case, there will again be no discrepancy between the position of the *New Theory* and that of the *Principles*. Berkeley's reworking of optics in terms of the tangible, so far from being an admission of materialism, can be seen as an important step in his ultimate argument against materialism.[12]

The Nature of Berkeley's Immaterialism

The discrepancies between the *New Theory* and the *Principles* are taken to be serious to the extent that the position of the *New Theory* is thought to be at variance with the immaterialism of the *Principles*. But these arguments which see the differences between the *New Theory* and the *Principles* as serious depend upon a particular version of what it is for a theory to be materialistic. The important difference between the *Principles* and the *New Theory* is that in the *Principles*, all sensible qualities, including tangible qualities, are mind-dependent, whereas in the *New Theory*, while visual qualities are said to be mind-dependent, the contrary is assumed for tangible qualities. If what makes a theory materialistic has to do solely with the status of sensible qualities, their mind-dependence or their mind-independence, then the judgment that the *New Theory* is materialist or more materialist than the *Principles* makes sense. It also requires us, however, to identify materialism as a thesis about sensible qualities. On this view, a theory is materialist to the extent that it takes sensible qualities to exist without the mind, whereas a theory will be imma-

11. I have written more about these issues in "Corpuscles, Mechanism and Essentialism in Berkeley and Locke," *Journal of the History of Philosophy*, forthcoming.

12. See Turbayne, "The Relationship between Berkeley's Science and His Metaphysics."

terialist if it takes sensible qualities to exist in the mind. The difference between materialism and immaterialism therefore would come down to a dispute about the ideality of sensible qualities. In fact, this way of looking at things identifies immaterialism and idealism.

I have argued, however, that it is not very helpful to look at the *New Theory* as primarily addressed to the ideality of the visuals. Such a view at the very least fails to do justice to the position against which Berkeley is arguing. The geometric optics Berkeley is trying to refute also holds that what we see is mind-dependent and that its character reflects the nature of the sensory systems by means of which we receive sensations. This is not, therefore, a part of the position he is trying to combat, and indeed, to combat this position would seem to require defending the claim that seeing is nonmental, or that it is not something that happens to perceivers, or that it is independent of any visual apparatus. What Berkeley is arguing against is a thesis about visual representation. He wants to argue that when what we see represents spatial properties like distance or size, this is not because what we see exhibits necessary or resembling connections with some mind-independent object. Instead what we see represents other sensory qualities, in this case, tangible qualities. What we see takes on a tangible meaning. The argument of the *New Theory* does not concern the status of visual qualities themselves but is instead about the way in which visual qualities represent something nonvisual.

Exactly the same can be said for the thesis of the *Principles*. The *Principles*, like the *New Theory*, can be seen as primarily concerned with the nature of what is represented by sensible qualities. In the *Principles*, Berkeley's primary focus is the notion of material substance, with the claim that what our ideas represent is a mind-independent material substance, on which they depend and with which they are held to be necessarily connected. His arguments are directed to showing that it makes no sense to suppose ideas can represent something nonideational and that no sense can be made of the claim ideas depend upon or are necessarily connected with something nonmental. Berkeley's target will again be, as it was in the *New Theory*, those theories, such as those of Descartes and Malebranche, which make important use of the notion of material substance. It is certainly not a characteristic of these theories to hold that *any* sensible qualities have a mind-independent existence. That which exists outside the mind is nonsensory in nature. Therefore the thesis to be contrasted with materialism is not a thesis about the ideality of the sensory, since this claim is also a feature of the materialism against which Berkeley is arguing. The materialism against which Berkeley directed his arguments is a thesis about what ideas represent: it claims what our ideas stand for is a mind-independent material substance. Immaterialism, the thesis Berkeley contrasts with materialism, is similarly a theory about what ideas represent. Considered negatively, it rejects the concept of material substance as useless and incoherent; considered positively, it claims that ideas represent only other ideas or, as Berkeley puts it in the passage from *Three Dialogues* quoted above, "the more a man knows of the connexion of ideas, the more he is said to know of the nature of things." Considered positively, immaterialism is the thesis that natural knowledge is a matter of learning to read the universal language of nature.

Many discussions of Berkeley's theory as it is found in the *Principles* and *Three Dialogues* tend to overemphasize the importance of the subjectivist or idealist claim that sensible objects exist only in the mind and, indeed, often treat this principle as if it represented the sum total of the positive position Berkeley is putting in place of materialism. Berkeley is taken to be arguing that materialists are committed to two incompatible claims: that we are aware of public objects, and that we are aware of private or inner states. Berkeley's position is then taken to be that since the view that we are aware of public objects, when these are identified with material substance, is incoherent, then the truth of the matter must be that all we are aware of are inner states or ideas and hence that things are to be identified with ideas.[13] The view that sensible objects exist only in the mind, that in sense perception what the perceiver immediately perceives are ideas of sense, though certainly an important premise

13. See, for example, Tipton, *Berkeley: The Philosophy of Immaterialism*, chap. 2: "Berkeley and Common Sense." Tipton's reaction to this line of argument is also typical. Whatever the merits of Berkeley's rejection of matter, he argues, does not make any more credible the claim that things are ideas. Therefore a more reasonable response to Berkeley's arguments against representational realism would be to explore the possibilities of direct realism.

in his argument, is not, however, by any means the whole of Berkeley's argument. It comes down to the relatively uncontroversial claim that sensations are ways of experiencing which happen to beings with minds capable of having experiences and which therefore are dependent upon the various sensory apparatus by means of which such beings have experience. There are clearly no ways of experiencing or perceiving in the absence of perceivers; hence sensations or sensible objects do not exist without the mind. What are immediately perceived are ideas of sense.

This claim, in and of itself, aids in the formulation of a problem but is not itself the solution to a problem. This is true for Berkeley as much as it is for his materialist rivals. Sensations, the various ways in which we experience, are proper to and dependent upon the various mechanisms of our different sense modalities. Visual sensations, the way things look to perceivers, are dependent upon our visual apparatus; hence the way things look are limited to lights and colors. But perceivers generally are said to see considerably more than light and color. They are said to see distance or to see tables, trees, or coaches. It is when Berkeley's idealism or subjectivism is stressed that he is sometimes said to deny we actually perceive distance or perceive physical objects like tables, trees, and coaches. But in fact his position is the same as that of the materialist, namely, we don't immediately perceive distance or tables or trees or coaches by sight. Berkeley however links the uncontroversial and shared thesis of idealism with his own solution to the problem, immaterialism, by means of the claim that sensations cannot represent something mind-independent either by resemblance or through necessary connections. When we explain how it is that we perceive distance or perceive a coach, we are explaining how it is that, on the basis of what we immediately perceive, we are able to acquire perceptual knowledge of what is not immediately available to a particular sense modality or modalities. The problem of perceptual knowledge generally—how it is that we, through the differing sensations of our different sense organs, gain knowledge of things that possess many different kinds of qualities—constitutes, in the *Principles* and *Three Dialogues*, the analog of the problem of visual space perception in the *New Theory*. Immaterialism is, then, first and foremost concerned with what we mediately or indirectly perceive. It says that what we mediately perceive, on the basis of what we immediately perceive, can be shown to be not representations of mind-independent material substance but other ideas of sense.

When Berkeley's various claims are distinguished in this manner, into what I have been calling an idealist component concerning the nature of sensory experience and an immaterialist component concerning the nature of sensory representation, then they do not conflict with common sense in quite the way that has often been alleged. The idealist component does not in and of itself conflict with common sense, inasmuch as any account of perception will have to reflect the fact that the way in which we perceive what exists is a subjective experience of the perceiver. Berkeley's views are taken to be noncommonsensical when it is supposed that claims he is actually making about the objects of immediate perception, claims expressing his idealism, are instead claims about the objects of mediate perception, which are properly expressions of immaterialism. In these circumstances, when Berkeley says, for example, that we never see tables or trees, then he is taken to be saying, contrary to common sense, that we never perceive physical objects like tables or trees. George Pitcher, for example, attributes to Berkeley the view that "we never perceive so-called physical objects—e.g. never see such things as tulips or rocks; we perceive *only* things that exist in our own minds" (p. 144). Pitcher's gloss on the noncommonsensical claim that we never perceive physical objects runs together issues of immediate and mediate perception. It is quite true that, according to Berkeley, we can't immediately see tulips or rocks, but the reason why we can't immediately see tulips or rocks is not that we see only our own ideas. It is because all that we can immediately see are those visual qualities apparent from our current vantage point, and being a tulip or a rock involves considerably more than just this narrow range of visual qualities but includes many other qualities that are proper to other sense organs and hence can't be *seen* at all. Berkeley is in fact not saying we can't perceive or mediately see tulips or rocks. We do indeed identify elements of our immediate visual experience as being tulips or rocks. We take what we are immediately perceiving to represent the stable range of qualities that has been in our experience reliably associated together and that we call being a tulip or being a rock.

Berkeley's theory is that our ability to perceive physical objects like tulips or rocks can be explained without asserting that what we immediately perceive represents anything other than our own ideas. When, by virtue of what we immediately see, we perceive a tulip or rock, we have not linked up a mental representation with something nonmental, as the single unchangeable nature of tulips or rocks, but instead, in Berkeley's way of looking at things, have assigned a meaning to what we see. Since this meaning is expressed in terms of ways in which the tulip or the rock is experienced, Berkeley can claim that, according to his theory, the tulip or the rock that is mediately perceived, that is the object of representation, has those sensible properties assigned by common sense to tulips and rocks, for the tulip is experienced as red and the rock as solid. Berkeley is not flouting common sense and denying we can perceive physical objects. What he is asserting is that understanding the nature of physical objects is a matter of understanding how our ideas or ways of experiencing go together. On this account, then, what is needed if we are to know about or to enlarge our knowledge of tulips or rocks is not, as the materialists would have it, distorted by the senses but is, instead, provided by the senses.

Immaterialism and Skepticism

An important element of Berkeley's project in *Principles* and *Three Dialogues* is the refutation of skepticism.[14] Immaterialism is advanced by Berkeley as a way of overcoming the skepticism he regarded as endemic to materialism. Berkeley's judgment that materialism is conducive to skepticism can be usefully connected, as Richard Popkin has shown, with skeptically motivated arguments such as those of Bayle.[15] Bayle argues that the same sorts of reasoning that Cartesians use to induce a skepticism of the senses can be extended to the knowledge of mind-independent matter, which the

14. Richard Popkin, in his "Berkeley and Pyrrhronism," *Review of Metaphysics*, 5 (1951–52) 223–46, has called attention to the fact that the full titles of both books mention that the work in question will be concerned with skepticism.

15. Popkin, "Berkeley and Pyrrhonism." Popkin refers particularly to Bayle's articles on Pyrrho of Elis and on Zeno the Eleatic in his *Dictionnaire historique et critique*.

Cartesians hoped would replace a sense-based epistemology. Berkeley's deployment of arguments of this sort has been regarded as among the most successful parts of his project. But they have not on the whole been regarded as effective arguments in favor of immaterialism. Berkeley is taken to be arguing that since skepticism is the natural result of holding that ideas stand for mind-independent things, the way out of skepticism is to identify things with ideas.[16] This is again to identify immaterialism and idealism. And it is widely held that, no matter how serious the problems Berkeley has raised for materialism, they cannot justify the move to idealism.

I have argued, however, that the identification of things with ideas is not in fact the whole of Berkeley's solution to the problem of skepticism. Berkeley's theory of space perception, as developed in the *New Theory*, shows he has more to offer as a refutation of skepticism than the identification of things with ideas. The *New Theory* offers a positive program to account for space perception, a program intended to replace the account on which many of the Cartesian arguments for skepticism of the senses are based. In the *New Theory*, we are shown that a theory in which ideas represent other ideas is one that can explain how we do successfully perceive spatial properties and is a better theory than the geometric theory, which purports to show that the perception by the senses of spatial properties is necessarily imperfect. Similar sorts of claims can be made for Berkeley's treatment of the wider issues of *Principles* and *Three Dialogues*. His concern is to show the weakness of materialist theories, and here, part of his argument is that the various demonstrations of skepticism with regard to the senses can be extended to our alleged ability to grasp the nature of mind-independent matter. But the rest of his argument defends the senses as a means of natural knowledge. Berkeley's claim that our ideas represent other ideas and his exploitation of the analogy with language form the underpinnings of this positive program.

Berkeley's demonstration that materialism can lead to skepticism

16. See, for example, Tipton: "Berkeley's solution to the problem is in essence staggeringly simple. He holds that if the root cause of skepticism is to be found in the supposition that real things are distinct from ideas, opposition to skepticism must rest on the contrary supposition, viz, that ideas and things are to be identified" (*Berkeley: The Philosophy of Immaterialism*, p. 53).

is often thought to depend upon a rather general argument exploiting the problems of the veil-of-perception: if all we are aware of is our own ideas, and things are other than ideas, then claims made about things cannot be substantiated by means of what we are aware of, our ideas.[17] Berkeley himself, however, tends to link the arguments that lead to a skepticism of the senses with a particular picture of what the world behind the veil-of-perception is like. For example, in *PHK* 102, he offers the following diagnosis of the arguments of those skeptics who hold that we are misled by our senses: "One great inducement to our pronouncing our selves ignorant of the nature of things, is the current opinion that every thing includes within it self the cause of its properties: or that there is in each object an inward essence, which is the source whence its discernible qualities flow, and whereon they depend." The sorts of doubts about the senses with which Berkeley is concerned are those which claim the mind-dependent or body-dependent deliverances of the senses are inadequate to inform us of the real nature of things. They concern the nature of sensory representation. He has in mind arguments that depend upon a particular picture of what the world out there to be known must be like, namely, that each thing has its own "inward essence," which our senses are unable to register. This is the general sort of picture Berkeley has refuted in the *New Theory*, where he has shown it is incorrect to regard sensory spatial information as a misleading representation of the real or absolute mind-independent space on which the representation is held to depend. His arguments in the *New Theory* required a reconceptualization of what the space to be known is like, in which Berkeley encouraged us to see the development of spatial knowledge as based on the perception of regularities or generalities among experiences.

17. Tipton, for example, writes: "The point can be made in various ways. To tie it in with our discussion of the Causal Theory: if the external objects are causes and we are aware only of effects, scepticism about the existence of the external objects is bound to follow. To tie it in with our discussion of qualities: if it is allowed that there are external objects and that these have qualities, then if we are aware only of ideas, we can never know which qualities should be attributed to a given object. And to tie it in with Malebranche: if sense experience informs us only of the visible or intelligible then skepticism concerning the existence of the material is again unavoidable" (*Berkeley: The Philosophy of Immaterialism*, p. 52). In each case, the point, according to Tipton, concerns the general difficulty of moving from ideas to conclusions which are about nonideas.

Berkeley's refutation of skepticism in his later works in based on a similar reconception of what is to be known. For example, in *PHK* 105, he writes:

> If therefore we consider the difference there is betwixt natural philosophers and other men, with regard to their knowledge of the phenomena, we shall find it consists, not in an exacter knowledge of the efficient cause that produces them, for that can be no other than the *will of a spirit*, but only in a greater largeness of comprehension, whereby analogies, harmonies, and agreements are discovered in the works of Nature, and the particular effects explained, that is, reduced to general rules.

Berkeley's case is that it is possible to overcome a skepticism engendered by the claim that our senses systematically mislead us with respect to the nature of things, if we recognize the incoherence of an attempt to ground a knowledge of nature on the uncovering of inward essences on which all other qualities depend. Instead, he recommends we understand the search for natural knowledge to be a matter of including events and processes under laws of increasing generality.[18] Berkeley doesn't simply justify the reliance on sensory information by arguing that, by these means, we avoid the skeptical difficulties that lie on the other side of the veil-of-perception. Rather, his view is that by relying on sensory information alone, we are able to acquire an understanding of natural phenomena that avoids the incoherences of a theory based on essences. As in the *New Theory*, Berkeley's ultimate justification for his approach is that it will work. We can therefore regard the project of understanding and explaining natural phenomena as one for which our human mental capacities are eminently suited.

Berkeley is trying to reject the view there are limitations on what we can come to know which result from the sensory basis of human knowledge. Thus his project is to defend the senses as adequate to reveal to us the nature of things. This claim, that there are no inherent limitations placed on sensory knowledge, is not, of course, the same as a claim that we do, or ever will, know everything there is

18. I have argued further for this claim in my "Corpuscles, Mechanism and Essentialism in Berkeley and Locke."

to know about the nature of things. Understanding natural events and processes is a matter of subsuming them under laws of increasing generality, which laws are derived from phenomena by observing analogies and similarities. There is nothing about our mental capacities that prevents us from carrying out this task. On the other hand, it is clear there are many regularities that have and will remain unnoticed by us. There are many pitfalls, moreover, that must be avoided, in particular that of overgeneralizing. Berkeley's example here is that of attraction. Just because there is a class of bodies that are attracted to one another does not mean there is not another class of bodies that repel one another (*PHK* 106). It is important not to leap to the conclusion that an observed regularity is essential or will hold universally. Berkeley can maintain there is a great deal that can stand in the way of our ever knowing everything there is to know about the nature of things without supposing that we are held back by inherent limitations on our mental faculties.

The Interpretation of Later Works in Light of the New Theory

I have suggested the relationship between the *New Theory* and Berkeley's later writings ought to be seen as close, so close that a proper understanding of the *New Theory* will provide a guide for the interpretation of such works as *Principles* and *Three Dialogues*. If we look for ways in which the *New Theory* can be taken to be a special case of the general line of argument to be maintained in *Principles* and *Three Dialogues*, then several points of resemblance emerge, which, I have argued, can serve as the basis for a way of reading the later works, a reading that leaves Berkeley considerably less at variance with common sense than is often thought to be the case.

The most important issue that is highlighted by a comparison between the *New Theory* and the later works concerns the kind of question Berkeley is trying to answer. In the *New Theory*, Berkeley is looking for a solution to a problem. He is seeking to discover how we perceive spatial properties by sight. What makes this a problem is the content of what we see. Visual ideas are inappropriate to convey spatial information, and yet we perceive space by sight. Thus the focus of the question Berkeley is asking is about the nature of visual representation: How is it that ideas of visual qualities can stand for or represent nonvisual qualities? This question presupposes an account of the nature of sensory objects, in particular that it is possible to identify a specific range of qualities proper to a given sense whose apprehension will be dependent upon the proper functioning of that sense. This thesis, I have argued, was not regarded by Berkeley as controversial, nor should it be given a controversial reading. What ought to be taken to be Berkeley's unique contribution in the *New Theory* is his account of visual representation. Similarly, then, we can see Berkeley in the *Principles* and *Three Dialogues* as asking a question about the nature of sensory representation and not a question about the nature of sensory objects. The focus of Berkeley's immaterialism, and what will distinguish it from materialism, will be an account of the ways in which sensory ideas represent rather than a claim that sensory objects exist only "in the mind." As in the *New Theory*, this latter claim should be regarded as a part of a relatively uncontroversial posing of a problem that the theory of sensory representation is trying to solve. And as in the *New Theory*, Berkeley's question assumes that our senses do, in fact, represent; that just as we do perceive distance, so we do perceive physical objects.

When Berkeley's problem is conceived in this manner, then certain positive aspects of his program swing to the fore. The presence of this positive program in the *New Theory* is quite apparent. Berkeley spends a good deal of time showing how it is we perceive spatial properties by sight and detailing the nature of the spatial properties we perceive. The solution to his problem requires him to establish that visual space perception can best be understood when we recognize that visual signs take on a spatial meaning when they are found to coexist with the very different ideas we acquire tangibly. The entire thrust of the work is positive, and by the time Berkeley announces his conclusion that vision is the language of the Author of nature, he has already spent much time developing this claim. In *Principles* and *Three Dialogues*, Berkeley allocates his time somewhat differently, and much more of these works is spent demonstrating the failures of the prevailing materialist theory. The tone of these books is much more negative. Nevertheless, I have argued that the theory Berkeley is defending in these two works, immaterialism, ought to be understood as having positive as well as negative import; in fact, the same positive import as his theory of visual representation. Although Berkeley is arguing negatively for the incoherence of a theory that takes the deliverances of the senses to represent mind-independent matter, he is also arguing positively that sensory representation is languagelike, that sensory ideas are signs that stand for other ideas of sense. As in the *New Theory*, Berkeley is trying to establish that our senses work, that they are adequate to inform us of the nature of things. And, as in the *New Theory*, this claim requires a reconceptualization of the nature of things our ideas represent. In the *New Theory*, we learn about our spatial surroundings when we learn to associate our visual ideas with conceptually unrelated and nonresembling ideas of touch. So in the *Principles* and *Three Dialogues*, learning more about the nature of things is a matter of expanding the associations among our ideas. What we are learning about are those generalizations we call the laws of nature.

The *New Theory* has a very specific target against which it is directed, what Berkeley calls the geometric theory of vision. I have argued that it is best to see this target as psychological in nature: Berkeley is arguing against those theorists who hold that vision is, as a psychological process, best understood as a kind of reconstructive geometry. The theories of vision that most clearly exemplify this psychological approach to vision are that of Descartes and, even more clearly, that of Malebranche. The *New Theory* argues that a Cartesian theory ought to be replaced by a theory based on the association of ideas. Berkeley's claim that what we need to be able to do if we are to understand natural phenomena is to recognize regularities, or laws of nature, while a novelty in the context of a theory of vision, is not unprecedented as a claim about natural knowledge generally. In the first edition of the *Principles*, Berkeley refers to Newton's *Principia* as "the best grammar of the kind we are speaking of" (*PHK* 110), even though he moves on immediately to express his reservations about some of the notions Newton employs. One way, therefore, to look at what Berkeley is doing in the *New Theory* as well as later works is to see him as encouraging us to replace a science based on the Cartesian model, which is essentialist in nature, with one, based on his own language model, which is lawlike. Thus the debate Berkeley could see himself as engaged in is one that pits the Cartesian view of natural science, with its tendency toward skepticism, against a more Newtonian view, which emphasizes the importance of general laws derived from phenomena.

Bibliography

Texts and Translations

Berkeley, George. *The Works of George Berkeley, Bishop of Cloyne* (9 vols). Edited by A. A. Luce and T. E. Jessop. Edinburgh: Thomas Nelson and Sons, 1948–57.

Descartes, René. *Discourse on Method, Optics, Geometry, Meterology.* Translated by Paul J. Olscamp. Indianapolis: Bobbs-Merrill, 1965.

Descartes, René. *Le Monde, ou Traité de la lumière*. Translation and introduction by Michael Sean Mahoney. New York: Arabis Books, 1979.

Descartes, René. *The Philosophical Writings of Descartes*. Translated by John Cottingham, Robert Stoothoff, and Dugald Murdoch. Cambridge: Cambridge University Press, 1985.

Descartes, René. *Principles of Philosophy*. Translated by Valentine Rodger Miller and Reese P. Miller. Dordrecht: D. Reidel, 1983.

Descartes, René. *Treatise on Man*. Translation and commentary by Thomas Steele Hall. Cambridge, Mass.: Harvard University Press, 1972.

Hooke, Robert. *Micrographia*. London, 1665; reprint Codicote, Herts.: J. Cramer, Weinheim, Welson, and Wesley, and New York: Hafter Publishing, 1961.

Locke, John. *An Essay Concerning Human Understanding*. Edited by Peter H. Nidditch. Oxford: Clarendon Press, 1975.

Malebranche, Nicolas. *Dialogues on Metaphysics*. Translation and introduction by Willis Doney. New York: Arabis Books, 1980.

Malebranche, Nicolas. *Father Malebranche his treatise concerning the search after truth*. Translated by T. Taylor. London: W. Bowyer for Thomas Bennet and T. Leigh and D. Midwinter, 1700.

Malebranche, Nicolas. "Reponse à Regis." In *Pièces jointes et écrits divers*. Edited by Pierre Costabel, Armand Cuvillier, and Andre Robinet. Paris: J. Vrin, 1960.

Malebranche, Nicolas. *The Search after Truth* and *Elucidations of the Search after Truth*. Translated by Thomas P. Lennon and Paul J. Olscamp. Columbus: Ohio State University Press, 1980.

Molyneux, William. *Dioptrica nova*. London: Benj. Tooke, 1792.

General Works

Abbott, Thomas K. *Sight and Touch: An Attempt to Disprove the Received (or Berkeleian) Theory of Vision*. London: Longman, Green, Longman, Roberts and Green, 1894.

Arbini, Ronald. "Did Descartes Have a Philosophical Theory of Sense Perception?" *Journal of the History of Ideas*, 21 (1983) 317–37.

Armstrong, D. M. *Berkeley's Theory of Vision*. Melbourne: Melbourne University Press, 1960.

Armstrong, D. M. "Discussion: Berkeley's *New Theory of Vision*." *Journal of the History of Ideas*, 17 (1956) 127–29.

Atherton, Margaret. "Berkeley's Anti-Abstractionism." In *Essays on the Philosophy of George Berkeley*, ed. Ernest Sosa, pp. 45–60. Dordrecht: D. Reidel, 1987.

Atherton, Margaret. "Corpuscles, Mechanism and Essentialism in Berkeley and Locke." *Journal of the History of Philosophy*, forthcoming.

Bailey, Samuel. *A Review of Berkeley's Theory of Vision*. London, 1842.

Berman, David. "Berkeley and the Moon Illusions." *Revue Internationale de la Philosophie*, 154 (1985) 215–22.

Bolton, Martha Brandt. "Berkeley's Objections to Abstract Ideas and Unconceived Objects." In *Essays on the Philosophy of George Berkeley*, ed. Ernest Sosa, pp. 61–81. Dordrecht: D. Reidel, 1987.

Brook, Richard J. *Berkeley's Philosophy of Science*. The Hague: Martinus Nijhoff, 1973.

Brykman, Genevieve. "Berkeley: Sa lecture de Malebranche à travers le dictionnaire de Bayle." *Revue Internationale de Philosophie*, 114 (1975) 496–514.

Brykman, Genevieve. "Microscopes and Philosophical Method in Berkeley." In *Berkeley: Critical and Interpretive Essays*, ed. C. M. Turbayne, pp. 69–82. Minneapolis: University of Minnesota Press, 1982.

Brykman, Genevieve. *Berkeley, philosophie et apologétique*. Paris: J. Vrin, 1984.

Cummins, Phillip D. "On the Status of the Visuals in Berkeley's *New Theory of Vision*." In *Essays on the Philosophy of George Berkeley*, ed. Ernest Sosa, pp. 165–94. Dordrecht: D. Reidel, 1987.

Davis, John W. "The Molyneux Problem." *Journal of the History of Ideas*, 21 (1960) 392–408.

Dicker, Georges. "The Concept of Immediate Perception in Berkeley's Immaterialism." In *Berkeley: Critical and Interpretive Essays*, ed. C. M. Turbayne, pp. 48–66. Minneapolis: University of Minnesota Press, 1982.

Donagan, Alan. "Berkeley's Theory of the Immediate Objects of Vision." In *Studies in Perception*, ed. Peter K. Machamer and Robert G. Turnbull, pp. 312–35. Columbus: Ohio State University Press, 1978.

Epstein, William. "Historical Introduction to the Constancies." In *Stability and Constancy in Visual Perception*, ed. Epstein. New York: John Wiley, 1977.

Furlong, E. J. "Berkeley and the 'Knot about Inverted Images.'" *Australasian Journal of Philosophy*, 41 (1963) 306–16.

Garber, Daniel. "Locke, Berkeley and Corpuscular Skepticism." In *Berkeley: Critical and Interpretive Essays*, ed. C. M. Turbayne, pp. 174–93. Minneapolis: University of Minnesota Press, 1982.

Gibson, J. J. *The Perception of the Visual World*. Boston: Houghton Mifflin, 1950.

Hatfield, Gary, and William Epstein. "The Sensory Core and the Medieval Foundations of Early Modern Perceptual Theory." *Isis*, 70 (1979) 363–84.

Hicks, G. Dawes. *Berkeley*. New York: Russell and Russell, 1968.

Kaufman, Lloyd, and Irvin Rock. "The Moon Illusion." In *Perception: Mechanisms and Models*. San Francisco: W. H. Freeman, 1972.

Kline, A. David. "Berkeley, Pitcher, and Distance Perception." *International Studies in Philosophy*, 12 (1980) 1–8.

Lennon, Thoman M. "Berkeley and the Ineffable." *Synthese*, 75 (1988) 231–50.

Lindberg, David C. *Theories of Vision from Al-Kindi to Kepler*. Chicago: University of Chicago Press, 1976.

Lindberg, David C., ed. *John Pecham and the Science of Optics*. Madison: University of Wisconsin Press, 1970.

Loeb, Louis. *From Descartes to Hume: Continental Metaphysics and the Development of Modern Philosophy*. Ithaca: Cornell University Press, 1981.

Luce, A. A. *Berkeley and Malebranche: A Study in the Origins of Berkeley's Thought*. Oxford: Oxford University Press, 1934.

McCracken, Charles. *Malebranche and British Philosophy*. Oxford: Clarendon Press, 1983.

Maull, Nancy. "Cartesian Optics and the Geometrization of Nature." In *Descartes: Philosophy, Mathematics and Physics*, ed. Stephen Gaukroger, pp. 23–40. Brighton, Sussex: Harvester Press, 1980.

Mill, John Stuart. "Bailey on Berkeley's Theory of Vision." In *Dissertations and Discussions*, 2:84–119. New York: Haskell House, 1973.

Pappas, George. "Berkeley, Perception and Common Sense." In *Berkeley: Critical and Interpretive Essays*, ed. C. M. Turbayne, pp. 3–21. Minneapolis: University of Minnesota Press, 1982.

Pappas, George. "Abstract Ideas and the '*esse* is *percipi*' Thesis." *Hermathena*, 139 (1985) 47–62.

Pappas, George. "Berkeley and Immediate Perception." In *Essays on the Philosophy of George Berkeley*, ed. Ernest Sosa, pp. 195–213. Dordrecht: D. Reidel, 1987.

Park, Desirée. "Locke and Berkeley on the Molyneux Problem." *Journal of the History of Ideas*, 30 (1969), 253–60.

Pitcher, George. *Berkeley*. London: Routledge & Kegan Paul, 1977.

Pitcher, George. "Berkeley and the Perception of Objects." *Journal of the History of Philosophy*, 24 (1986), 99–105.

Popkin, Richard. "Berkeley and Pyrrhonism." *Review of Metaphysics*, 5 (1951–52) 223–46.

Raynor, David. "'Minima Sensibilia' in Berkeley and Hume." *Dialogue*, 19 (1980) 196–200.

Ritchie, A. D. *George Berkeley: A Reappraisal*. Edited by G. E. Davie. Manchester: Manchester University Press, 1967.

Rock, Irvin. *An Introduction to Perception*. New York: Macmillan, 1975.

Schwartz, Robert. "Seeing Distance from a Berkeleian Perspective." Unpublished.

Senden, Marius von. *Space and Sight*. Glencoe, Ill.: Free Press, 1960.

Sosa, Ernest, ed. *Essays on the Philosophy of George Berkeley*. Dordrecht: D. Reidel, 1987.

Thrane, Gary. "Berkeley's 'Proper Object of Vision.'" *Journal of the History of Ideas*, 38 (1977) 243–60.

Tipton, I. C. *Berkeley: The Philosophy of Immaterialism*. London: Methuen, 1974.

Turbayne, C. M. "Berkeley and Molyneux on Retinal Images." *Journal of the History of Ideas*, 16 (1955) 339–55.

Turbayne, C. M. "The Influence of Berkeley's Science on His Metaphysics." *Philosophical and Phenomenological Research*, 16 (1956) 476–87.

Turbayne, C. M. "Editor's Commentary." In George Berkeley, *Works on Vision*. Indianapolis: Bobbs-Merrill, 1963.

Turbayne, C. M. *The Myth of Metaphor*. Rev. ed. Columbia, S.C.: University of South Carolina Press, 1971.

Turbayne, C. M., ed. *Berkeley: Critical and Interpretive Essays*. Minneapolis: University of Minnesota Press, 1982.

Vesey, G. N. A. "Berkeley and the Man Born Blind." *Proceedings of the Aristotelian Society*, 61 (1960–61) 189–206.

Warnock, G. J. *Berkeley*. London: Penguin, 1953.

White, A. R. "The Ambiguity of Berkeley's 'Without the Mind'." *Hermathena*, 83 (1954) 55–65.

Wilson, Catherine. "Visual Surface and Visual Symbol: The Microscope and the Occult in Early Modern Science." *Journal of the History of Ideas*, 49 (1988) 85–108.

Wilson, Margaret D. "Berkeley and the Essences of the Corpuscularians." In *Essays on Berkeley*, ed. John Foster and Howard Robinson, pp. 131–47. Oxford: Oxford University Press, 1985.

Winkler, Kenneth. "Berkeley on Abstract Ideas." *Archiv für Geschichte der Philosophie*, 65 (1985) 63–88.

Winkler, Kenneth. *Berkeley: An Interpretation*. Oxford: Oxford University Press, 1989.

Yolton, John W. *Perceptual Acquaintance from Descartes to Reid*. Minneapolis: University of Minnesota Press, 1984.

Index

Library of Congress Cataloging-in-Publication Data

Atherton, Margaret.
Berkeley's revolution in vision / Margaret Atherton.
p. cm.
Includes bibliographical references.
ISBN 0-8014-2358-9 (alk. paper)
1. Berkeley, George, 1685–1753. Essay toward a new theory of vision. 2. Vision.
3. Immaterialism (Philosophy) 4. Berkeley, George, 1685–1753—Contributions in theory of vision. I. Title.
B1339.A72 1990
121′ .3—dc20 89-49632